Vivek Adhikari
Mahendrasinh Dabhi

# Book on Nano-emulsions against lesser grain borer on stored wheat

Vivek Adhikari
Mahendrasinh Dabhi

# Book on Nano-emulsions against lesser grain borer on stored wheat

## Nano-emulsions of botanicals against lesser grain borer

LAP LAMBERT Academic Publishing

**Imprint**



Cover image: www.ingimage.com

Publisher:
LAP LAMBERT Academic Publishing
is a trademark of
Dodo Books Indian Ocean Ltd. and OmniScriptum S.R.L publishing group

120 High Road, East Finchley, London, N2 9ED, United Kingdom
Str. Armeneasca 28/1, office 1, Chisinau MD-2012, Republic of Moldova, Europe
Managing Directors: Ieva Konstantinova, Victoria Ursu
info@omniscriptum.com

Printed at: see last page
**ISBN: 978-620-8-44334-4**

**Acknowledgement**

I found indeed proud to mention that my Lord Swaminarayan Bhagvan , Guru Hari Shri Param Pujay Bharamswarup Pramukh Swamimaharaj and Pragat Guru Hari Shri Param Pujay Bharamswarup Mahant Swamimaharaj has provided me strength for assessment of this subject books in my field of specialization. My Guru has also provided me inspiration for the book writing and also gave me his continuous blessings during the entire period of book writing. I would like to express my special thanks to my field of teaching that have given me an opportunity to stand for the same. I am also very thankful to my wife Shobhna and my son Aksharraj for providing me spare time from their lovely life.

**Mahendrasinh V Dabhi**

A journey is easier when you travel together." Interdependence, indeed, proves to be more valuable than independence. With profound gratitude and heartfelt appreciation, I present this thesis, which stands as the culmination of two years of unwavering support from numerous personalities. This moment allows me the privilege to extend my sincere thanks to all those who have played an indispensable role in shaping this work.

First and foremost, I express my deepest sense of gratitude to my major advisor, Dr. M. V. Dabhi, Assistant Professor at S.M.C. Polytechnic in Agriculture, B.A. College of Agriculture, Anand Agricultural University, Anand. His invaluable guidance, constant inspiration, and meticulous supervision have been the pillars that guided me through this journey. His unwavering faith in research excellence and insistence on producing high-quality work has left a lasting impression on me, making me profoundly fortunate to have worked under his mentorship.

I extend my sincere thanks to Dr. D. B. Sisodya, Professor & Head, Department of Agricultural Entomology, whose humane attitude, sustained encouragement, and critical suggestions have been instrumental in refining my work.

I am equally grateful to my minor advisor, Dr. R. K. Gangwar, Associate Research Scientist at the Main Rice Research Station, AAU, Nawagam, and the members of my advisory committee, Dr. A. D. Kalola, Associate Professor & Head, Department of Agricultural Statistics, and Dr. R. K. Thumar, Associate Professor & Head, AICNP on Ornithology, for their kind counsel and valuable insights during my study and research.

Special thanks to Dr. M. C. Mankad, Prof. N. A. Bhatt, Vaibhav sir, Dr. Kalyanrao Patil and Dr. H. Pandey for their valuable suggestions and constructive critiques, which have enriched the quality of this work.

I am deeply appreciative of Dr. Y. M. Shukla, Principal, B.A. College of Agriculture, AAU, Anand, and the University authorities for granting me the opportunity to pursue higher studies at such a prestigious institute.

I would like to extend my sincere appreciation to Trusha Ma'am, Kinjal Behen, Rutul Sir, and Muskan Behen from the Department of Nanotechnology for their invaluable help and guidance in preparing nano-emulsions during my research.

I am immensely grateful to my senior guide mate, Anjalibehen Chauhan, for her generous sharing of experience, encouragement, and valuable knowledge. Her contributions have been instrumental in shaping the outcome of this work.

My heartfelt thanks go out to my batchmates Anita, Ganesh (Vlogger), Manvi, Bhautik, Divyesh, Ajay, Dimple, Shabnam, Divya and seniors Pavan sir, Ravi bhai, Jay bhai, Ajay bhai, Rashmit bhai, Atul bhai, Abhishek bhai, Senthil sir, Bhavik bhai, Birla sir, Suwalal sir, Lokeshwaran sir and Harsh bhai, as well as friends Harshit, Abhishek, Prakash, Rakesh, Subham and 'ICAR Buddies'. Their unreserved support, camaraderie, and enduring companionship have made this academic journey memorable and fulfilling.

I extend my thanks to the juniors from the department, Deepak, Soham, Jayant, Henil, Khush, Daxa, Vruti, Pruvi, Jayalekshmi, Khyati, and Srushti, for their constant and invaluable help.

I express my heartfelt gratitude to my juniors Chandan, Deepak, and Prajwal for their invaluable help during my busy times. I also want to thank Adi, Sanjit, Rajesh and Ahwan for the positive vibes they brought from the very beginning.

I wish to express my sincere appreciation to every staff member of the Department of Agril. Entomology (Babu kaka, Lala bhai, Savan bhai, Rekha behen, Minaxi maam, and Dharmesh bhai) and Regional Research Station (Pavan sir, Balwantbhai, Ramilabehen, Piyush bhai, Bhavesh bhai and Soundaryabehen) for creating a lively environment and assisting me throughout my research.

I am grateful to ICAR for granting me the opportunity to pursue my master's degree at AAU, Anand. My sincere thanks go to the entire team of NAHEP-CAAST, AAU, Anand, especially Dr. R.S. Pundir, Rucha Ma'am, and Jignesh Bhai, for their invaluable support. I am equally thankful to AIT, Thailand, particularly Dr. Avishek Datta, Dr. Hayat Ullah, Dr. Sushil Himanshu and Arindam Sir, for providing me with the opportunity for a valuable foreign exposure visit.

I extend my heartfelt thanks to Adi and Zayra for their unwavering support and understanding, which provided me with the opportunity to engage in part-time teaching at Dharmanos High School,USA. This experience not only helped me meet my financial requirements during my studies but also enriched my teaching skills and provided valuable experience in the field of education.

I am sincerely grateful to my relatives, neighbors, Parumala ammachis, Mannar friends and Asha teacher for their kind prayers and unwavering moral support. Their encouragement has been a source of strength throughout my journey.

None of this would have been possible without the unwavering support and love from my parents, Ramakrishna Adhikari and Devi Adhikari, who have been the bedrock of my life. Their blessings, patience, and overwhelming encouragement have been my guiding light throughout this journey. To my loving brother Vinod, I am forever indebted for his unwavering moral support during all my ups and downs. My entire family has been a constant source of inspiration and moral support throughout my academic career, and for that, I am immensely grateful. Their sacrifices have made my dreams come true.

I am deeply aware of the oceans of guidance, support, and encouragement that I have received from countless sources. I take this opportunity to extend my profound sense of gratitude and heartfelt appreciation to all those who have contributed to my growth during this academic pursuit.

Lastly, I bow to the almighty for guiding me through life, providing me with opportunities, and fulfilling my dreams.

**Vivek Adhikari**

## Contents

## Chapter-I

## Introduction

Wheat (*Triticum aestivum* Linnaeus) belongs to Poaceae family which is the second maximum produced grain among cereals after maize (Anonymous, 2014). It is one of the most important cereals grown globally and one of the critical staple meal of almost 2.5 billion population. It is called as "king of cereals". Globally, it is the main source of plant-based protein (12 %) in the human diet. Wheat contains about 70 % carbohydrates, 12 % protein, 2.7 % minerals, 2 % dietary fiber and 12 % moisture content. Wheat is used in the form of flour to make chapattis, semolina and pasta products. It is also used for making bread, cookies, biscuits, cracks, pasta, dahlia, maida, vermicelli *etc.*

In 2022-23, 781.31 million metric tons of wheat was produced worldwide (Shahbandeh, 2023). India is the second largest wheat producer in the world. In 2022-2023, the crop was cultivated on 30.54 million hectares (14 % of the global area) to produce the highest ever production of 112 million metric tons of wheat (13.64 % of world production) with average record productivity of 3371 kg/ha. In the financial year 2022-23, over four million metric tons of wheat was produced in 1.2 million hectares in Gujarat with productivity of 3172.41 kg per hectare (Anonymous, 2022). The wheat production target is projected to be around 140 million tons by 2050, taking into account the growing demand for consumption and trade in India (Anonymous, 2020).

Storage pests can cause post-harvest losses estimated up to 9 % and 20 % in developed and developing countries, respectively (Phillips and Throne, 2010). In India, post-harvest losses caused by unscientific storage, insects, rodents, microorganisms *etc.* account for about 10 per cent of all food grains. Storage wheat is heavily infested by various insect pests *viz.*, lesser grain borer (*Rhyzopertha dominica* Fabricius), rice weevil (*Sitophilus oryzae* Linnaeus), red rust flour beetle (*Tribolium casteneum* Herbst.), granary weevil (*Sitophilus granarius* Linnaeus), maize weevil (*Sitophilus zeamais* Motschulsky), angoumois grain moth (*Sitotroga cerealella* Olivier) and rice moth (*Corcyra cephalonica* Stainton) (Ileke and Oni, 2011). Among these insect pests, *R. dominica* (Coleoptera: Bostrichidae) and *S. oryzae* (Coleoptera: Curculionidae) are the primary pests of stored wheat and feed internally as well as externally on both whole kernel grains and cereal products (Burges, 2008 and Mark *et al.*, 2010).

Lesser grain borer, *R. dominica* is a pest which starts its infestation from field and also damage stored grains and this may cause economic damage during storage (Adedire, 2001). The adults are powerful fliers and migrate from one godown to another, causing a fresh infestation. When the infestation is severe, the adults produce a considerable amount of frass, spoiling more than what it eats. The flour, so produced, serves as nourishment for the young grubs until they are ready to bore into the grain. They not only reduce the quantity but also deteriorate the quality of grain and its products (Atwal, 1994).

This borer is the only species in the genus *Rhyzopertha* and belongs to subfamily Dinoderinae under the family Bostrichidae. It is said to have originated in India (Pruthi and Singh, 1950). Its narrow cylindrical form and tiny size helps to distinguish it from other grain beetles. In females, the fifth sternal segment is pale yellow, but it is brown in males. Eggs are typically deposited around the embryo of grain, either singly or in a row, where young larvae can easily enter the grain. The eggs hatch and the larvae tunnel within a grain, where it completes its development and emerge from the grain through a big exit hole. It is active in a wide range of humidity and may feed on grains with less than 9 per cent moisture content, under which other stored grain insect pests normally do not survive or stop feeding (Edde, 2012). The knowledge on identification, habit and life cycle of this pest will help to formulate appropriate management strategies.

Stored grain pests can cause significant economic losses to farmers and food processing industries, as well as pose a threat to food security. They can easily infest and damage stored wheat, reducing the quality and quantity of grain and its products. In order to prevent these losses, it is essential to implement effective management strategies that can control the spread of pests and minimize their impact. Management of stored grain pests is always evolving and requires timely observation of pest biology and its behavior, evaluation of efficacy of current management practices, resistance development, testing and release of novel management strategies which can give better results with less environmental impact and cost.

Traditionally, the use of synthetic pesticides has been the primary method for managing storage pests. However, the overuse of pesticides has led to the development of resistance in pests, as well as posed risks to human health and the environment. Therefore, there is a growing need to find alternative and eco-friendly pest management strategies that can minimize the use of synthetic pesticides. (Sarkar *et al.*, 2021).

One promising alternative is the use of non-edible oils to control storage pests. These oils are derived from various plant sources and have been found to be effective in controlling a wide range of insect pests, including *R. dominica* and *S. oryzae*. Oils of neem, pongamia, castor, rape seed, sesame, coconut mustard, tobacco seed, groundnut and cotton seed have shown repellent activities against stored grain pest (Kher, 2006). The secondary plant metabolites present in oils may act as kairomones, allomones, stimulants or deterents of feeding and oviposition and as antifeedants, insecticides and insect hormone mimics (Talukder, 2006). In addition, the use of nano-emulsions of these oils has been shown to enhance their effectiveness, even at lower concentrations.

Neem oil is obtained from neem seed kernels of neem tree, *Azadirachta indica* Juss. It contains at least 100 biologically active compounds. Among them, the main components are the triterpenes known as limonoids, the most important being azadirachtin, which appears to exert 90 % of the effects on most pests and has multiple mode of actions (Campos *et al.*, 2016). Castor oil is a triglyceride containing about 90 % of fatty acids, containing main toxicants ricin and ricinolein (Perdomo *et al.*, 2013). Karanj oil is extracted from seeds of karanj tree, contains tannins, trypsin inhibitors and the toxin karanjin (Natanam *et al.*, 1989). Eucalyptus oil is obtained hydro distillation of leaves from eucalyptus tree belonging to the myrtaceae family has 700 species distributed throughout the world (Tandon *et al.*, 2008) and main components present are α-pinene, 1,8-cineole and transpinocarveol (Khemira *et al.*, 2012).

Nano-emulsions are stable dispersions of oil droplets in water, with droplet sizes typically in the range of 20-300 nanometers (Fernandez *et al.*, 2004). Nano-emulsions are typically prepared using high energy equipment's such as ultrasonicators, microfluidizers and high-pressure homogenizers. These droplets have a large surface area which allowed greater contact with the insect pests, leading to increased toxicity. In addition, nano-emulsions have been found to be more stable and longer-lasting than traditional emulsions, making them a more effective and sustainable pest management tool (Golden *et al.*, 2018).

The use of non-edible oils and their nano-emulsions has several advantages over traditional pesticides, which are environment friendly and pose minimal risks to human health and the ecosystem. It is also readily available and can be easily synthesized which made it cost-effective and accessible to farmers. Finally, these products have been found to be effective against a wide range of insect pests, making them a versatile and sustainable pest management strategy.

Another potential method for managing insect pests is through the use of stored grain's natural resistance. In order to effectively manage storage pest infestation, choosing resistant varieties to cultivate for long-term storage can serve as a valuable supplement to other management activities. It is important to note that storage grain pests display preferences for certain varieties of grains and may find others less suitable for feeding, oviposition, and development. By comparing the biophysical parameters of different cultivars, we can identify and avoid susceptible varieties, potentially reducing the risk of pest infestation. Grains of different varieties vary in their morphological, biophysical, and biochemical properties, and pests tend to infest grains with smooth surfaces and thin seed coats. By gaining a better understanding of pest behaviour, we can make informed decisions and select the appropriate cultivars for long-term storage of wheat (Sarin and Sharma, 1983).

Using of non-edible oils to control storage pest is gaining popularity because they are eco-friendly in nature. Nano-emulsions perform better than normal oils even in very lower concentrations. Identification and improvisation of potential non-edible oils and their nano-emulsions will definitely bring up increased use of oils into management practices for storage pests and hence pesticide load in food chain could also be reduced. Choice of resistant varieties along with management practices may help to enhance safe storage of grains.

Keeping in view the above facts, present investigations were carried out with the following objectives.

**Objectives**

- To study the bionomics of lesser grain borer on stored wheat
- To screen different varieties of wheat for susceptibility to lesser grain borer
- To evaluate different non-edible oils and its nano-emulsions as grain protectants against lesser grain borer on stored wheat

## Chapter- II

## Review of Literature

The current study intends to look into bionomics of lesser grain borer, *Rhyzopertha dominica* on stored wheat, its susceptibility to wheat cultivars, and the use of non-edible oils and nano-emulsions as potential grain protectants. Thus, providing useful insights into cultivar selection, sustainable pest management measures in grain storage facilities.

This review of literature will delve into previous research on the bionomics, host susceptibility, and the use of non-edible oils and nano-emulsions for pest management in order to establish a comprehensive understanding of the subject matter.

### 2.1 BIONOMICS OF *R. dominica* ON STORED WHEAT

Halstead (1963) reported that it is possible to distinguish between the sexes of lesser grain borer during the pupal stage. Genitalia of the females are divergent, three-segmented and protuberant whereas, those of the males are convergent, two-segmented and protuberant.

Distribution pattern of lesser grain borer on wheat grains filed in hollow metallic tower showed that significantly more damage was found in the central portions when compared to sides as lesser grain borers were congregated at the central portion most of the times. However, no significant damage was found below 50.8 cm (Keever, 1983).

Bashir (2000) found the existence of a distinct transverse, punctuate groove on the fifth abdominal sternite of the male *R. dominica* from Pakistan, which is never present in females.

Mumm *et al.* (2001) examined the thermal biology of *R. dominica* by offering adult beetles a range of temperatures on a laboratory thermal gradient. After seven days, it was found that 88% of the beetles were located within a temperature range of 29.0 to 34.0 °C, which was also the preferred temperature range for oviposition.

According to Campbell (2002) *R. dominica* was fully active at temperatures $\geq 14.6 \pm 0.1$°Con a laboratory thermal gradient. After 24 hours, it was incapable of surviving temperatures ≥46.5± 0.1°C. The preferred temperature for survival and oviposition of *R. dominica* was 29.4± 0.1°C.

*R. dominica* achieved its maximum reproductive success on dry wheat grains. The adult beetle is 2 to 3 mm long and 0.8 to 1 mm wide. Fresh body weight of adult ranged from 0.99 to 1.38 mg, whereas mean fresh body weight was about 1.20 mg (Edde and Phillips, 2006).

Bio-ecological studies of *R. dominica* carried out by Kumawat (2007) revealed that maximum and minimum fecundity of 307 and 60 eggs per female occurred at a temperature and relative humidity of 30 ± 1°C&75 ± 5% R.H. and 40±1°C &65±5% R.H., respectively. The incubation and pupal periods were 5.3 to 9.4 days and 5.00to 6.7 days, respectively. Female longevity was 30.9 days at 25±1°C temperature and 75±5% R.H. and decreased at higher temperatures. Maximum (74.0%) and minimum (2.0%) survival of eggs occurred at 30±1°C&75±5% R.H. and 40±1°C & 65±5% R.H., respectively. Moreover, the growth index of the lesser grain borer was found to be high (3.6) at the temperature of 30±1°C.

Eggs of *R. dominica* are deposited in clusters on grain or singly among the frass. The egg is opaque, whitish in colour with a waxy appearance when freshly laidbut,later on takes on a pinkish colour (Kucerova and Stejskal, 2008).

Klys (2008)conducted experiments on population development and feeding activity of *R. dominica* under three temperature variants28, 22 and 31°C. Population development was inhibited below 22 °C and feeding was highest at 31°C.

According to Mason and McDonough (2012), larvae of lesser grain borer are white and C-shaped with four to five larval instars. Both the larvae and adults are voracious feeders. Adults usually remain within the kernel for a few days prior to emergence. Mated females start ovipositing about two weeks later and continue for about four months.

Naik*et al.* (2016) studied the influence of food sources*viz.*, whole wheat, paddy, sorghum and broken wheat on developmental period of *R. dominica*. The developmental period of egg, larva and pupa was4.73, 28.20&5.00 and 6.38, 33.40&7.00 days in wheat and paddy, respectively. However, the total development period recorded was 37.93 and 46.78 days on wheat and paddy, respectively.

Chintala *et al.* (2017) recorded the incubation period of 5 to 6 days,larval period of 12 - 19 days, pupal period of 3 - 5 days, Male to female sex ratio of 1.09: 1.0 and viability of eggs was 90 % in case of *R. dominica.* Adults were dark reddish brown in colour and 3 mm long having antennae with three segmented club. The head was masked by pitted prothoracic notum. The elytra consisted of distinct rows of pits along the length.

Kumar *et al.* (2017) studied comparative biology of *R. dominica*in major cereals and observed that the males usually have the last exposed abdominal tergum uniformly brown and that of the female is usually pale yellow.

The seasonal variations in the biology of *R. dominica* were investigated by Ajaykumara *et al.* (2018) through laboratory experiments conducted during *kharif* 2014, *rabi*2014 and summer 2015. The results indicated a longer total grub period of 39.32 days during *kharif* and a shorter period of 31.28 days during summer. Male and female adult beetles lived longer in *kharif* with durations of 26.08 and 28.64 days, respectively and shorter in summer with durations of 19.52 and 21.88 days, respectively. The mean total life cycle duration was 60.96and 49.76 days during *kharif* and summer, respectively. The incubation period, duration of each in starand total life cycle duration varied significantly between the seasons.

Morphometric studies of *R. dominica* revealed that the average length and breadth of egg were 0.56 ± 0.06 mm and 0.23 ± 0.04 mm, respectively. The average length and breadth of first instar larva were 0.78 ± 0.04 mm and 0.20 ± 0.04 mm, respectively. The average length and breadth of second instar larva were 1.09 ± 0.06 mm and 0.30 ± 0.03 mm, respectively. The average length and breadth of third instar larva were 2.08 ± 0.08 mm and 0.40 ± 0.04 mm, respectively. The average length and breadth of fourth instar larvae were 3.10 ± 0.05 mm and 0.61 ± 0.08 mm, respectively (Chintala and Virani, 2018).

Female remains underneath the male while mating and thus reported a method to separate male and female without intensive observation on limited sexual dimorphism. Average duration for completion of life cycle was 57.33, 48.33, 40.33 and 34.00 days at 25 ± 2, 27 ± 2, 30 ± 2 and 35 ± 2 °C temperature and 60 ± 5, 70 ± 5, 75 ± 5 and 85 ± 5 percent relative humidity (Deshwal *et al.*, 2018).

Win and Rolania (2020) observed that the sex ratio (male: female) was 0.84:1.19 and 0.90 : 1.10 in winter and rainy season, respectively which indicated the predominance of females over males. The fecundity was ranged from 123 to 215 (170.8 ± 16.08) and 285 to 398 (306.8 ± 16.36) eggs in winter and rainy season, respectively.

Scheff *et al.* (2021) observed greater presence of lesser grain borer during warmer months and found a significant correlation between temperature and lesser grain borer activity. Pheromones were more attractive to lesser grain borers than natural stored wheat.

Studies conducted by Awadalla *et al.* (2023) on biological aspects of the lesser grain borer revealed that development period was 25.40 ± 0.75 days. The number of progenies emerged were 46.80 ± 0.73 and 76.20 ± 4.27individuals for first and second generations, respectively.

Studies on antenna of lesser grain borer by Seada and Hamza (2023) indicated that antenna consisted of a scape, a pedicel, and eight segmented fagellum divided into narrow proximal funicle with five segments and three wide distal clubs.

## 2.2 SCREENINGOF DIFFERENT VARIETIES OF WHEAT FOR SUSCEPTIBILITY TO LESSER GRAINBORER*R. dominica*

Rao and Wilbur(1972) recorded a weight loss of 9.5 % due to twenty days of larval feeding. Weight loss due to adult feeding were 19.4, 12.0, 9.5, and 6.5 % per kernel during $1^{st}$, $2^{nd}$, $3^{rd}$ and $4^{rth}$ week, respectively.

Mcgaughey *et al.* (1990) evaluated wheat varieties for susceptibility to lesser grain borer and found that seed hardness did not had a significant impact on the reproduction of lesser grain borers.

Evaluation of susceptibility of wheat cultivars infested by *R. dominica* by revealed that hardness was not significantly related with reproduction of lesser grain borer (Toews *et al.*, 2000).

Afshan (2005)examined the population development and damage caused by the lesser grain borer, *R. dominica*, on wheat grains (Abadgar variety).The population buildup of *R. dominica* was significantly higher during the period of high temperatures and decreased gradually with reduction in temperature. The grain weight loss was recorded to be 3.73% after a month of introduction of the pest and it increased to 25.55% after five months. Population buildup and grain weight loss were positively correlated with temperature and relative humidity.

The relative resistance of wheat varieties against the lesser grain borer *R. dominica*, was investigated by Khokhar (2005)and it was found that population increased on all varieties before gradually declining. Variety abadgar showed the lowest population pressure (282.71) proving to be the most resistant, followed by mehran (722.83) and TD1 (731.37). On the other hand, varieties such as sarsabz, kiran, TJ85, anmol, momal, and TJ83 were found to be susceptible against the pest. Additionally, the study found a positive and highly significant relationship between the population of lesser grain borer and percent grain weight loss.

Studies carried out on post-harvest resistance of wheat varieties against lesser grain borer by Broughton and Dunkel (2006) revealed that the overall protein content and endosperm hardness did not have a significant relationship with feeding and frass production.

Ten different varieties of wheat were screened for their susceptibility to *R. dominica*. GW 173 and GW 190 were found resistant. The number of adults emerged and germination loss at the end of 5 month after infestation in case of varieties LOK 1 and GW 496 were 101.4 & 108.78 and 70.25 % & 56.97 %, respectively. The variety LOK 1 was found highly susceptible followed by moderately susceptible GW496 (Kher, 2006).

Sayed *et al.* (2006) evaluated the resistance of twelve wheat varieties against the lesser grain borer and kaphra beetle. Their findings revealed that mehran 89 exhibited the highest resistance against both of these pests, as indicated by the lowest population growth and least weight loss. On the other hand, TJ-0787 and sarsabz demonstrated the lowest resistance, based on highest population growth and weight loss.

Khanzada (2007) evaluated the susceptibility of various wheat varieties to lesser grain borer and found that TD1, ZA77, LYP73 and AS2003 were the most susceptible with the highest weight loss and number of adults emerged. Faisalabad85 was found to be resistant to lesser grain borer which could be due to its harder seed coat.

Based on fecundity, hatching, pupation, adult mergence, development periods, number of generations, adult longevity, $F_1$ offspring and susceptibility index, study indicated that cultivars K9107and K9162 were preferred however cultivars K68, K8027, K9465, PBW343 were less preferred foods for *R. dominica* (Meenakshi& Srivastava,2010).

Pal*et al.* (2011) investigated the effect of different levels of *R. dominica* infestation on deterioration of wheat seeds and found that population growth and per cent seed infestation were directly proportional to the level of *R. dominica* infestation. It was observed that there wassignificant reduction in germination when more than two pairs of adults were released.

Obretenchev and Staneva, (2012) tested the resistance of thirty soft wheat varieties against *R. dominica* was tested under laboratory conditions. The results showed that variety VIII-30-a-2-4 was the most resistant, followed by slavijanka, with the average number of developed insects being 66.7±4.87 and 84.7±6.34, respectively. Conversely, mirijana and prelom were the most susceptible varietieswith an average number of developed insects of 609.7±15.14 and 693.4±19.37, respectively.

According to Bhargava & Akhter (2014), susceptibility of fifteen varieties of wheat was evaluated against lesser grain borer. Lowest and highest number of eggs were laid on Raj1482 and Raj3077, respectively. Larval and pupal development periods were found to be maximum in Raj1482 and minimum in C306. Grain damage and weight loss also varied among the different varieties which varied from 17.32 to 45.79% and 6.15 to 18.50%, respectively.

Eyidozehi *et al.* (2014)tested eight different lines and cultivars of wheatfor their resistance to lesser grain borer and observed that there was no significant difference in the characteristics of the wheat varieties for damage and feeding by the lesser grain borer. However, the morvarid variety showed the least reduction in viability (10.34%), seed weight (13.07%) and seed impurity (12.84%). Darya had the highest percentage decrease in seed weight (21.97%) and seed impurity (21.94%). The pastor variety had the highest percentage decrease in viability (27.33%).

Chintala (2015) screened seven varieties of wheat against *R. dominica* and found significantly minimum (6.50 %) grain damage, population build up (4.00

adults), frass weight (0.88 g), minimum period of life cycle (29.69 days) in the variety GW-366. The population growth of this pest in the varieties LOK1 and GW496 were found to be 5.87 and 10.13, respectively. The chronological order of superiority of the varieties based on overall performance was GW366 > GW322 > LOK1 > GW273 > GW496 > GW503 > GW190.

In free choice test, conducted by Korawar (2019) based on adult orientation and colonization, five varieties *viz*., NIAW3581, NIAW3643, NIAW3270, NIAW3523 and NIAW317 were considered as susceptible, while four varieties MACS6222, NIAW343, NIAW1415 and NIAW3592 were proved as tolerant against *R. dominica*.

Evaluation of wheat varieties against *R. dominica* by Mehta (2020) showed that HPW155 had maximum adult emergence (111.67), susceptibility index (11.69), orientation preference (24.44) and shortest developmental period (40.33), while, HPW349 had minimum adult emergence (54.00), susceptibility index (7.40), orientation preference (7.67) and longest developmental period (54.00) in *R. dominica*.

Shukla *et al*. (2020) reported that, among different wheat varieties, the lowest larval period of *R. dominica* (27.50 days) was found in HI 7747, and also larva could not complete its development even up to 48 days and it was dead before pupation. The highest larval period (33.44 days) was observed in UPT72294 followed by K 65 (32.42 days) and TL174 (31.54 days).

Among different varieties, HD 2967 was found highly susceptible on account of maximum adult emergence, grain damage and weight loss, while the variety PDW 291 was the least preferred by *R. dominica* (Singh and Sharma, 2021).

Out of fifteen varieties screened by Kumawat *et al*. (2022), Raj4037, HI153, MP1234, HI8627, GW366, MP3288 and HI1500 were found moderately resistant while Anuradha, KRL14, HI1544, Lok1, MPO1203, HD2932 and MPO1215 were susceptible and HI 1567 was highly susceptible to *R. dominica*.

Gami *et al*. (2022) tested ten wheat varieties/genotypes (GADW1, GADW3, N820, N817, N814, N810, N807, N813, N809, N811) against *R. dominica* on stored wheat and found that the varieties N813 and N809 were resistant which showed minimal seed damage of 25.09% and 25.94%, minimal weight loss of 11.77% and

14.64% and a higher germination rate of 58.45% and 53.64%, while, minimal adult emergence was 61.33 and 44.47, respectively, after 180 days of storage. GADW3 and N811 cultivars were considered susceptible which had higher seed damage of 44.44% and 40.44%, major weight loss of 35.79% and 33.72%, minimal germination of 14.30% and 22.69%, and higher population growth of 390.33 and 367.00, respectively.

Baghla *et al.* (2023) screened different wheat varieties against lesser grain borer and results showed that Lok1, GW366, and GW173 had the highest percent grain damage, while HI1544 and HI8489D showed the least damage. HI1544 was found to be highly resistant with a minimum weight loss of 2.16%. On the other hand, Lok1 was preferred by *R. dominica* with a maximum weight loss of 15.40%. The population buildup was observed to be highest in Lok1 (42.5) followed by GW366 (35.50) and GW173 (30.50) and least in HI1544 (12) after 120 days of storage.

## 2.3 EVALUATION OF DIFFERENT NON-EDIBLE OILS AND ITS NANO-EMULSIONS AS GRAIN PROTECTANTS AGAINST LESSER GRAIN BORER, *R. dominica*ON STORED WHEAT

According to Klein and Hebbe (1995), germination of oil treated wheat seeds was reduced by 20-50 % and time to 50 % emergence was increased by two to five days.

Exposure of adults of both beetle species(*Cryptolestes pusillus*Schonherr and *R. dominica*) to maize and sorghum grains treated with 10 mL/kg of the different oils induced 100% mortality within 24 h. Oil treatments did not affect the germination or water absorption compared with untreated grains (Obeng- Ofori, 1995).

According to El-Sheikh*et al.* (2005),exposure of lesser grain borer adults to $LC_{50}$ values of essential oils extracted from peppermint, marjorana, lavender, sweet basil, spearmintand artemisia stimulated acetylcholinesterase, glutamic transaminase, and amylase.Additionally, the essential oils depleted the total proteins and total carbohydrates in body tissues, but increased the concentrations of total lipids in treated insects.

The chronological order of overall effectiveness of various plant oils as grain protectants of wheat against *R. dominica* was: neem > pongamia > castor> rape seed > sesame > coconut > mustard > tobacco seed > groundnut > cotton seed. Thus, among different oils tested at 0.5 % (v/w), non-edible oils were found more effective than edible oils (Kher, 2006).

Dinesh and Raj (2012) studied the efficacy of four vegetable oils *viz.*,neem (*Azadirachta indica*), eucalyptus (*Eucalyptus globulus*), sunflower(*Helianthus annus*) and castor (*Ricinus communis*) oils at 3 mL/kg which resulted in 100 per cent control and 1 mL/kg which resulted in 0.33, 0.46 and 0.55% control, respectively, against the pulse beetles in pigeonpea.

Nema and Bhargava (2012) evaluated the effectiveness of seven essential oils (neem, karanj, eucalyptus, castor, lemongrass, palmarosa and citronella) against *R. dominica* on wheat, applied at a rate of 1.0 mL/100 g seeds. Neem oil showed the most promising results, with a longer developmental period (55.11 days), lower adult emergence (22.22%), minimum grain damage (11.26%), and reduced weight loss (4.46%). The order of effectiveness of different oils was, neem > karanj > citronella > castor > eucalyptus > lemongrass > palmarosa.

The maximum protection against *R. dominica* was provided by neem oil and tamaria oil which resulted in emergence of only 9.3 adults followed by mustard oil (9.7), castor oil (12.7), karanj oil (16.00) and eucalyptus oil (16.30) at 1.0 per cent. Germination test of oil treated seeds after 270 days indicated no significant loss in germination.The maximum grain damage was recorded with eucalyptus oil at 0.1 per cent followed by karanj oil at 0.1 per cent and eucalyptus at 0.5 per cent (Kumawat and Naga, 2013).

Chintala (2015) reported that minimum adult emergence of*R. dominica* (2.26) was recorded in grains treated with neem oil followed by karanj oil (3.66) and castor oil (4.13). However, maximum germination percentage was recorded from neem oil (93.5 %) followed by karanj oil (90.25 %) and castor oil (89 %).

The advantages of nano particles include larger surface area, easy attachment of single and several different pesticide molecules and a reasonably fast mass transfer to the target insects' body and delayed loss in efficacy due to degradation. Nano particles having insecticidal properties can be used as an active pesticide agent or biopesticide (Elango *et al.*, 2016).

Study conducted by Sabbour and El-Aziz (2016) on nano oils against *S. granarius* under laboratory and storage conditions revealed that, a lower concentration of castor nano-emulsion results in higher adult mortality when compared to higher concentration of normal castor oil.

Singh *et al.* (2016) evaluated the efficacy of three indigenous plant oils, namely neem oil, eucalyptus oil, and castor oil, against the lesser grain borer at three different concentrations (0.10%, 0.15%, and 0.20% v/w) by directly mixing with wheat seeds. The results of the study revealed that neem oil at 0.20% concentration was the most effective treatment against *R. dominica* with the minimum adult emergence (22.58 adults/100g grain sample), grain damage (4.79%), weight loss (2.60%)and maximum inhibition rate (88.92%), followed by eucalyptus oil and castor oil. The effectiveness of the oils declined with extended storage periods. However, the three oils had no adverse effect on seed viability for up to 120 days after treatment.

According to Hashem*et al.* (2018), the development of nano-emulsions allows to improve the stability of Anise, *Pimpinella anisum* essential oil enhancing its efficacy against stored grain pests and contributing to reduce the use of harmful synthetic insecticides.

Adak *et al.* (2020) concluded that the formulated eucalyptus nano-emulsion were superior to eucalyptus oil and it can be recommended as an alternative to harmful chemical pesticides against stored grain pests. Based on centrifugation, thermodynamic stability, heating and cooling, freezing and thawing experiments, two combinations, 1:2 and 1:2.5 (EO: Tween 80) of 6% eucalyptus oil concentration were optimum to formulate nano-emulsions. The droplet sizes of nano-emulsions were 4.04 nm and 2.27 nm and poly dispersity index (PDI) were 0.37 and 0.77 and zeta potential were 6.20 mV and 7.69 mV, respectively for these two combinations.

Studies by El-Naby *et al.* (2020) revealed 12 % and 25 % reduction in germination of wheat seeds treated with castor oil nano-emulsion and castor oil, respectively.

Many case studies reported higher mortality by nano-emulsions against various insect pests as compared to commercial insecticides and crude oil, hence nano-emulsions may be used as an alternative policy for the control of numerous insect pests (Sharma *et al.*, 2020).

Chandel*et al.* (2021) evaluated the fumigant toxicity and repellent action of *Eucalyptus globulus* Labill essential oil against adults of *R. dominica.* The concentration of 50, 100, 150 and 200 μL/L resulted in 80, 83.3, 93.3 and 96.6% mortality, respectively, after 24 hours of exposure. At the concentrations 0.08, 0.16, 0.24, and 0.32 $\mu l/cm^2$ the repellent activities were found to be 42.5, 43.5, 47.5 and 62.5%, respectively after 36 hours of exposure.

Kalaf (2021) studied fatal, attractive and repellent effects of five botanical oils *viz.,Cinnamomum camphora, Aloe vera, Zea mays, Mentha suaveolens*, and *Peganum harmala* on *R. dominica* in wheat. The average mortality was 89.79, 96.53, 87.48, 87.58, and 92.82%, respectively, while all oils had an attractive effect. The least population build up was observed in treatment with *Aloe vera* oil with an average of 2.66 individuals compared to 61.3 in control.

Abouelatta*et al.* (2022) evaluated the insecticidal activities of tagette oil against *R. dominica* and results showed that tagette essential oile xhibited fumigant effect on *R. dominica* with $LC_{50}$ value 0.26 and 0.17 $mg/cm^3$ after 3 and 6 h, respectively.Contact effect was also exhibited with $LC_{50}$ value of 820.7 and 665.5 mg/kgafter 24 and 48 h of exposure, respectively.

Ibrahim (2022) compared the toxicity index of free clove oil with clove oil nano capsules and revealed that nano capsules were 70% more toxic than free oil with$LC_{50}$ values 175.50 and 576.85 ppm, respectively.

The formulation of neem nano-biopesticide against *Tribolium castaneum* and *R.dominica*, as reported by Iqbal *etal.* (2022), resulted in a product with optimized physical characteristics, including an average particle size of 275.8 ± 0.95 nm, a PDI of 0.351 ± 0.002, and a zeta potential of –33 ± 0.90 mV. The nano-biopesticide also exhibited significantly higher mortality rates of 81 ± 3.04% and 84.97 ± 2.83% against *T. castaneum* and *R. dominica*, respectively, when compared to the crude extract. After three months of storage at 4 °C, a slight increase in particle size (from 275.8 ± 0.95 to 298.8 ± 1.00 nm) and PDI (from 0.351 ± 0.002 to 0.445 ± 0.02) was observed.

## Chapter- III

**Materials and Methods**

The lesser grain borer,*Rhyzopertha dominica* is a major pest of stored wheat and infested seeds are unfit for food and seed purpose. Thus, laboratory investigations on bionomics, varietal susceptibility, and the use of various non-edible oils and its nano-emulsions for their efficacy against *R.dominica* Fabricius infesting wheat under storage (Plate 3.1) were conducted in the seed laboratory of Regional Research Station, Anand Agricultural University, Anand during 2022-2023.

## MAINTENANCE OF LESSER GRAIN BORER CULTURE

For initiation of the culture, around 300 adults were collected from the local market of Anand. The adults of lesser grain borer, *R. dominica* were introduced into a plastic jar (15 cm x 8 cm)containing 1 kg wheat grains previously sterilized at a temperature of 55˚Cfor 4 hours in hot airoven [Plate 3.2(A)]. The jar was covered using muslin cloth tightened by rubber band to prevent the escape of adults. Adults of *R. dominica* were reared and maintained within the laboratory for population growth and development on wheat grains. The damaged grains were replaced by undamaged and sterilized grains at a monthly interval. Throughout the experiment, the culture was maintained in the laboratory [Plate 3.2(B)]. Thus, the adults of *R. dominica* obtained from the laboratory culture were used for further studies on bionomics, varietal susceptibility, evaluation of non-edible oils and its nano-emulsions as grain protectants.

## 3.1 BIONOMICS OF LESSER GRAIN BORER, *R. dominica* ON WHEAT

A laboratory study on bionomics of lesser grain borer, *R. dominica* infesting Lok 1 wheat was conducted at Regional Research Station, AAU, Anand with the following details.

### 3.1.1 Observations Recorded

- Egg Hatching (%)
- Incubation period (days)
- Larval period (days)
- Prepupal period (days)
- Pupal period (days)

- Adult longevity (days)
- Pre-oviposition period (days)
- Oviposition period(days)
- Post-oviposition period(days)
- Fecundity (No. of eggs/female)
- Sex ratio (Male: Female)
- Total life cycle (days)
- Morphometry of life stages

### 3.1.2 Methodology

The adults of lesser grain borer obtained from the culture maintained in laboratory was used to study bionomics of the pest.

#### 3.1.2.1 Egg

Five pairs of five to ten days old adults obtained from laboratory culture were released into 200 grams of wheat grains in a plastic jar (12 cm x 7 cm)for egg laying. In order to study the hatching percentage, the eggs laid by female adults were picked up with the help of fine camel hair brush and were transferred into a Petri plate(Diameter: 9 cm) [Plate 3.3(A)]. Twenty eggs were considered to calculate hatching percentage. The number of eggs hatched were recorded daily in the morning and evening till hatching. The eggs were considered as hatched when tiny larvae came out of it. Eggs were examined under the microscope to study their colour, shape and size. The length and breadth of eggs were measured using Zeiss stemi 2000-C stereozoom microscope and Axiovision Rel. 4.8 software [Plate 3.3(B)]. For determining incubation period, the eggs were examined from date of egg laying to the date of hatching.

#### 3.1.2.2 Larva

With a view to determine the number and duration of different larval instars, twenty freshly hatched larvae were reared individually by placing them on partially dissected wheat grains (LOK1 variety) in plastic vials (5 cm x 4.5 cm)[Plate 3.3(C)]. The grains were only partially dissected along ventral furrow or crease in such a way that it could be opened and closed like a book. The presence of larvae inside grain were indicated by dark tint in grain when observed under good light or artificial illumination [Plate 3.3(D)]. To observe the different larval stages, grains were opened and further dissected if larvae were mobilised to sides of grain. After observation of larval instars, the larvae were carefully placed back into dissected grains using camel hair brush. The observation on change of instar was done daily till they attained last instar. The exuviae of the larvae left out and change in size indicated the change of instars. The larval instar was studied for their colour, shape and size. The total larval period was calculated from egg hatching to the formation of prepupa.

#### 3.1.2.3 Pupa

The prepupal period was worked out from the day when last instar larvae became sluggish, stops feeding, slightly curved, contracted and started formation of pupae. Twenty larvae were observed for recording prepupal period. The pupal period was calculated from the day of formation of pupae to the day of emergence of adults. The pupae were collected from all the vials and were examined for their colour, shape, and size. The site of pupation was also observed.

#### 3.1.2.4 Adult

The emerged adults (n = 10) were paired for a day and then kept individually in vials (5 cm x 4.5 cm). The life span of each adult was recorded by counting the period from emergence of adult from pupa till death. Duration between emergence of adult to oviposition by female was considered as pre-oviposition period. Duration of egg laying was considered as oviposition period. To determine fecundity, eggs laid by females were collected and counted daily in the morning. The total number of eggs laid during life span of female adult was considered as fecundity. Duration between oviposition period to death of adult was considered as post-oviposition period. The adults were examined for their colour, shape and size. Male and female was differentiated by locating pale yellow colour in the abdominal segments of female adult. Number of males and females were counted to find sex ratio.

### 3.2 SCREENING OF DIFFERENT VARIETIES OF WHEAT FOR SUSCEPTIBILITY TO LESSER GRAIN BORER, *R. dominica*

All the varieties[Plate 3.4(A)]were procured from Regional Research Station, Anand Agricultural University, Anand; ARS, AAU, Arnej and ARS, AAU, Dhandhuka. The wheat varieties were evaluated based on population growth, germination loss and weight loss during storage. The laboratory experiments were carried out as per below mentioned methodology.

### 3.2.1 Details of Experiment

**Location** : Regional Research Station, Anand Agricultural University,Anand

**Year**: 2022-23

**Design** : Completely Randomized Design (CRD)

**Treatments** : 10

**Repetitions** : 3

**Treatment details**

| Tr. No. | Variety |
|---|---|
| $T_1$ | GW 513 |
| $T_2$ | GDW 1255 |
| $T_3$ | GADW 3 |
| $T_4$ | GW 451 |
| $T_5$ | GW 206 |
| $T_6$ | HI 1544 |
| $T_7$ | GW 499 |
| $T_8$ | GW 1 |
| $T_9$ | GW 496 |
| $T_{10}$ | LOK 1 |

### 3.2.2 Observations Recorded

- Population growth of *R. dominica*(No. of adults/100 g of grains)
- Weight loss (%)
- Germination loss (%)

- Seed size (mm)
- Seed weight (g)
- Seed hardness (N)

### 3.2.3 Evaluation Based on Population Growth of Lesser Grain Borer, *R. dominica*

To evaluate different wheat varieties based on population growth of *R. dominica,* three samples of wheat grains each of 100 g from each variety (one sample for one repetition) were filled in plastic vials(5.5 cm x 6 cm)individually. Five pairs of *R.dominica* adults (5 to 10 days old) were released into each vial for egg laying and each vial was covered with two-fold muslin cloth [Plate 3.4(B)]. The adult introduced for oviposition was discarded from each vial after 7 days of oviposition. The data on number of adults (live + dead) emerged was recorded monthly during six months of storage and was subjected to ANOVA after square root transformation.

### 3.2.4 Evaluation Based on Weight Loss of Grains

To evaluate wheat varieties based on weight loss of grains due to infestation by *R. dominica*, hundred grains were collected randomly from each sample and segregated into bored grain and germ eaten grain. The bored, germ eaten and 100 undamaged grains was weighed using electronic balance. Based on data, the per cent loss in weight was calculated by using the below mentioned formula given by Srivastava *et al.* (1973).

$$L = \frac{(W + G) - 100}{S(W_1 + G_1)}$$

Where,

L = Percent loss in weight

W = Percentage (by number) of bored grains

G = Percentage (by number) of germ eaten grains

S= Weight of 100 undamaged grains (g)

$W_1$ = Weight of bored grains (g)

$G_1$= Weight of germ eaten grains (g)

The data on per cent weight loss was subjected to ANOVA after arcsine transformation.

### 3.2.5 Evaluation Based on Reduction in Seed Germination

To evaluate wheat varieties based on germination loss due to infestation by *R. dominica*, laboratory germination test was conducted as per ISTA procedure by adopting 'between the paper method' (International Seed Testing Association, 2015)[Plate 3.5]. Hundred seeds in each repetition were taken at random from the treated seeds of each treatment and were placed uniformly on germination paper. The seed roll was kept in INDOSAW seed germinator at 20 ± 0.50 °C and 95 ± 1 per cent relative humidity. The final count for normal seedlings was carried out on eighth day of germination test and was expressed in percentage. The per cent loss in germination was worked out for each treatment based on the data on germination before and after 6 months of storage infestation. The data on per cent loss in germination was subjected to ANOVA after arcsine transformation.

### 3.2.6 Evaluation Based on Seed Size, Seed Weight and Seed Hardness of Grain

**Seedsize**

To record seed size, length and widthof wheat seeds were measured by INDOSAW seed image analyzer[Plate 3.6(A)] and was expressed in units of mm.

**Seed weight**

To record seed weight, a sample of hundred seeds was taken from each variety randomly and weighed out with the help of electronic digital scale.

**Seed hardness**

To record the seed hardness, three randomly selected grains wereanalyzed using a standard texture analyzer[Plate 3.6(B)]. Individual grainwas made to crack by applying pressure and the cracking point was noted down. The hardness was expressed in Newton.

Correlation between physical characters of the varieties and weight loss made by the pest was carried out.

### 3.2.7 Categorization of Varieties

The wheat varieties were grouped into six categories of susceptibility to *R. dominica viz.*, highly resistant, resistant, moderately resistant, moderately susceptible, susceptible and highly susceptible based on three parameters *viz.*, population growth, per cent loss in weight and germination loss. The mean value of individual variety (xi)was compared with mean value of all varieties ($\bar{x}$) and standard deviation (SD) following the scale adopted by Patel *et al.* (2002). Following scale was used categorizing different varieties (Table 3.1).

**Table 3.1: Scale of categorization of wheat varieties for resistance to *R. domnica***

| Category of resistance | Scale of resistance |
|---|---|
| Highly resistant | $\bar{x}_i \leq (\bar{x} - 2SD)$ |
| Resistant | $(\bar{x} - 2SD) < \bar{x}_i \leq (\bar{x} - SD)$ |
| Moderately resistant | $(\bar{x} - SD) < \bar{x}_i \leq \bar{x}$ |
| Moderately susceptible | $\bar{x} < \bar{x}_i \leq (\bar{x} + SD)$ |
| Susceptible | $(\bar{x} + SD) < \bar{x}_i \leq (\bar{x} + 2SD)$ |
| Highly susceptible | $\bar{x}_i > (\bar{x} + 2SD)$ |

## 3.3 EVALUATION OF DIFFERENT NON-EDIBLE OILS AND ITS NANO-EMULSIONS AS GRAIN PROTECTANTSAGAINST LESSER GRAIN BORER, *R. dominica* IN STORED WHEAT

Laboratory experiment was conducted to work out the efficacy of different non-edible oils and its nano-emulsions as seed protectants for safe storage of wheat seeds (LOK1) against *R. dominica*. 2 kg of wheat seeds was mixed with different plant oils and nano-emulsions [Plate 3.7(A)], and control was untreated. Pure oils were procured from local market and their nano-emulsions (6 %) [Plate 3.7(B)] were prepared using probe sonicator [Plate 3.7(C)]following standard procedures at the Department of Nanotechnology, AAU, Anand. These treated seeds were stored in plastic jars(15 cm x 8 cm). The mouth of the jars was covered with plastic lids. All the jars were kept in the laboratory [Plate3.8(A)] at room temperature and were utilized for further experimentations.

### 3.3.1 Details of Experiment

**Location** : Regional Research Station, Anand Agricultural University,Anand

**Year** : 2022-23

**Design** : Completely Randomized Design (CRD)

**Treatments** : 9

**Repetitions** : 3

**Treatment details**

| Tr. No. | Treatments | Dose (mL/100 g seeds) |
|---|---|---|
| $T_1$ | Castor oil 100 % | 1 |
| $T_2$ | Neem oil 100 % | 1 |
| $T_3$ | Karanj oil 100 % | 1 |
| $T_4$ | Eucalyptus oil 100 % | 1 |
| $T_5$ | Castor oil nano-emulsion 6 % | 1 |
| $T_6$ | Neem oil nano-emulsion 6 % | 1 |
| $T_7$ | Karanj oil nano-emulsion 6 % | 1 |
| $T_8$ | Eucalyptus oil nano-emulsion 6 % | 1 |
| $T_9$ | Control | - |

### 3.3.2 Observations Recorded

- Adult mortality (%)
- Population growth of *R. dominica* (No. of adults/100 g of grains)
- Germination loss (%)
- Hydrodynamic size of nano-emulsions (nm)

- Zeta potential of nano-emulsions (mV)
- Kilo count per second of nano-emulsions
- Polydispersity index of nano-emulsions

### 3.3.3 Preparation and Characterization of Nano-Emulsions

For preparation of nano-emulsions, 100 mL stock solution of tween-20 was prepared by mixing 12.5 mL tween-20 in 87.5 mL of water. 12.5 mL of this tween-20 stock solution was added drop by drop to 12.5 mL of oil in stirring condition on a magnetic stirrer. After it was properly mixed, 75 mL of Milli-Q water was added during stirring condition. Stirring was continued for another 20 minutes. The resultant emulsions obtained were kept for ultrasonication at a frequency of 20 kHz and amplitude of 60 μm in probesonicator for 5 minutes. The resultant nano-emulsions were allowed to cool down at room temperature. After diluting to 6 % concentration, it was characterize dusing particle size analyzer (Malvern Zetasizer ZS 90) [Plate 3.8(B)] and was used for further experiments.

### 3.3.4 Evaluation of Non-Edible Oils and Nano-Emulsions Based on Adult Mortality of *R. dominica*

To evaluate the non-edible oils and nano-emulsions based on adult mortality of *R. dominica*, a sample of 50 g of wheat seeds were placed in plastic vial(5 cm x4.5 cm) with three repetitions and twenty adults of *R. dominica* were released in each vial among all the treatments[Plate 3.8(C)]. Each vial was covered with two-fold muslin cloth to facilitate aeration and to prevent escape of adult. The observations on number of dead adults out of total adults was made after 7 days of adult release and per cent mortality was worked out. The experiment was repeated at 15 days interval following same methodology during storage period of 6 months. The periodical data on per cent mortality was corrected using Abbott's formula (Abbott, 1925) and was subjected to ANOVA after arcsine transformation.

$$P = \frac{P_1 - C}{100 - C} \times 100$$

Where,

$P$ = Corrected per cent mortality

$P_1$= Observed per cent mortality in treatment

$C$ = Per cent mortality in control

### 3.3.5 Evaluation of The Non-Edible Oils and Nano-Emulsions Based on Population Growth of *R. dominica*

To evaluate the non-edible oils and nano-emulsions based onpopulation growth of *R. dominica*, an experiment was carried out using same bulks of wheat grain under different treatments[Plate 3.8(D)]. The same methodology was followed as per heading 3.2.3. The data obtained was subjected to ANOVA after square root transformation.

### 3.3.6 Evaluation of The Non-Edible Oils and Nano-Emulsions Based on Wheat Seed Germination Loss

To evaluate the non-edible oils and nano-emulsions based on wheat seed germination loss, an experiment was carried out using same bulks of wheat grain under different treatments. The same methodology was followed as per heading 3.2.5. The data obtained was subjected to ANOVA after arcsine transformation.

## Chapter- IV

## Results and Discussion

The results of the present studies carried out on bionomics, varietal susceptibility, and the use of various non-edible oils and its nano-emulsions for their efficacy against *R. dominica* infesting wheat under storage are presented here under suitable sub-headings and discussed with relevant review related to the present findings.

### 4.1BIONOMICS OF *R. dominica* ON STORED WHEAT

Bionomics of lesser grain borer,*R. dominica*was carried out on LOK 1 variety of wheat at average temperature of 30 ± 2 °C and 75 ± 5 % R.H. The study was carried out at Seed Laboratory, Regional Research Station, AAU, Anand, during the month of February - March, 2023. The results of investigation on bionomics of lesser grain borer are depicted under Tables 4.1 & 4.2, and Plates 4.1 & 4.2.

#### 4.1.1 Egg

Study on egg stage of *R. dominica* was carried out using Zeiss stemi 2000-C stereozoom microscope at Department of Plant Biotechnology, AAU, Anand. Eggs were found laid mostly in group or singly among the mixture of debris and frass at bottom of the storage container and rarely singly on grain surface. The eggs were opaque, white in colour and oblong in shape with one end slightly pointed than the other [Plate 4.1(B)].Similar results were confirmed with Kucerova and Stejskal (2008) who reported that the eggs were opaque, whitish in colour with a waxy appearance and laid in clusters on grain or singly among the frass.

##### 4.1.1.1 Incubation period

The data from Table 4.2 indicated that incubation period ranged from 5 to 6 days with an average of 5.4 ± 0.5 days. The present findings were more or less similar to previous studies carried out by Chintala *et al.* (2017) and Kumawat (2007) who recorded the incubation period ranging from 5 to 6 days and 5.3 to 9.4days, respectively.

##### 4.1.1.2 Hatching percentage

The results presented in Table 4.2 revealed that 80 % of eggs were hatched and gave rise to first instar larvae. However, the present findings vary with Chintala *et al.* (2017) who reported hatching percentage of 90 % and it could be due to change in location and variety.

#### 4.1.1.3 Morphometry

The length of egg ranged from 0.46 to 0.55 mm with an average of 0.51 ± 0.04 mm and breadth of egg ranged from 0.18 to 0.23 mm with an average of 0.19 ± 0.01 mm respectively. Morphometric data investigated on eggs were found similar to the results from previous studies conducted by Chintala and Virani (2018) who reported that average length and breadth of egg was 0.56 ± 0.06 mm and 0.23 ± 0.04 mm, respectively.

### 4.1.2 Larva

Freshly hatched larvae were reared individually by placing them on partially dissected wheat grains. Grains were opened and dissected daily to observe larval instars. The exuviae of the larvae left out and change in size indicated the change of instars. Four larval instars were observed. All larval instars bored inside the grain and fed on endosperm and embryo portion until pupation. As a result of this, exuviae, frass and bored tunnel was seen inside grains. The total larval duration was varied from 20 to 29 days with an average of 24.35 ± 2.83 days. Findings from present studies are more or less similar to results from previous studies by Ajaykumara *et al.* (2018) who reported that average larval duration was31.28 days.

#### 4.1.2.1 First instar

The first instar larvae appeared tiny, translucent, whitish with hairy body and brownish head. The larvae were campodeiform in nature. Body segmentation was not fully visible. The length of first instar larvae varied from 0.68 to 0.75 mm with an average size of 0.71 ± 0.01 mm. The breadth of first instar larvae varied from 0.15 to 0.19 mm with an average size of 0.18 ± 0.00 mm. Findings from present studies are more or less similar to results from previous studies by Chintala and Virani (2018) who reported that average length and breadth of first instar larvae was 0.78 ± 0.04 mm and 0.20 ± 0.04 mm, respectively.

#### 4.1.2.2 Second instar

Second instar larvae were marked by an increased size, reduced translucency, whitish body with hairs. Improved visibility of abdominal segmentation was observed expect the posterior portion which remained voluminous and lustrous. The body of larvae were found slightly curved while resting. The minimum and maximum length of second instar larvae observed was 1.10 and 1.40 mm, respectively, with average value of 1.22 ± 0.06 mm. The minimum and maximum breadth of second instar larvae observed was 0.36 and 0.47 mm, respectively, with average value of 0.40 ± 0.02 mm. Findings from present studies are more or less similar to results from previous studies by Chintala and Virani (2018) who reported that the average length and breadth of second instar larvae were 1.09 ± 0.06 mm and 0.30 ± 0.03 mm, respectively.

#### 4.1.2.3 Third instar

Third instar larvae were scarabeiform (C- shaped), opaque, with white robust body and marked by clear body segmentation with hairs. Increased pigmentation of head and mandible were observed during this stage. The minimum and maximum length of third instar larvae observed was 1.98 and 2.34 mm, respectively, with average value of 2.18 ± 0.06 mm. The minimum and maximum breadth of second instar larvae was 0.56 and 0.75 mm, respectively, with average value of 0.63 ± 0.04 mm. Similar results were reported by Chintala and Virani (2018) who reported that the average length and breadth of third instar larvae were 2.08 ± 0.08 mm and 0.40 ± 0.04 mm, respectively.

#### 4.1.2.4 Fourth instar

Fourth instar larvae were scarabeiform (C- shaped), opaque, with white robust body and marked by clear body segmentation with hairs. Size and pigmentation of head and mandibles were enhanced during this stage. The minimum and maximum length of fourth instar larvae was measured as 2.58 and 2.72 mm, respectively, with average value of 2.68 ± 0.02 mm. The minimum and maximum breadth of second instar larvae observed was 0.89 and 1.03 mm, respectively, with average value of 0.98 ± 0.02 mm. Similar results were reported by Chintala and Virani (2018) who reported that the average length and breadth of fourth instar larvae were 3.10 ± 0.05 mm and 0.61 ± 0.08 mm, respectively

### 4.1.3 Prepupa and Pupa

Prepupal stage was marked by straightened and slightly shrunken body with presence of minor movement restricted to posterior abdominal portion. The minimum and maximum length of prepupae observed was 2.56 and 2.70 mm, respectively, with average value of 2.66 ± 0.02 mm. The minimum and maximum breadth of pupae were0.88 and 1.01 mm, respectively, with average value of 0.97± 0.01 mm. Prepupal duration lasted for one day after which body was transformed to pupal stage.

Pupae were exarate in nature. The pupation took place inside the seed at cavity made by larvae as a result of feeding activity. Initially, it was translucent and white in colour with hairs and pigmentation only on eyes, but later, light brown pigmentation was observed on all over pupal body. Presence of antennae was first observed during pupal stage. The minimum and maximum length of pupae observed were 2.52 and 2.61 mm, respectively, with average value of 2.52 ± 0.02 mm. The minimum and maximum breadth of pupae observed were 0.94 and 1.00 mm, respectively, with average value of 0.96 ± 0.01 mm. Pupal period varied between 3 to 6 days with average value of 3.85 ± 0.87. The results from current studies are nearly matching with the previous studies by Kumawat (2007), Naik *et al.* (2016) and Chintala *et al.* (2017) who reported pupal period ranging from 5 to 6.7 and 3 to 5 days, respectively.

#### 4.1.4 Adult

After completing pupal stage, fresh adult remained inside grain cavity for few days before boring out of grain. Such adults had light brown colour and were delicate compared to few days' old counterparts. The reason for adult to stay inside the grain for few days before emerging out could be due to a natural waiting period or acclimatization time in order to undergo complete sclerotization of body parts and mandibles so as to survive and feed efficiently in unpredictable external ambience. Adults were active fliers and were involved in boring grains, mating and dispersal.

##### 4.1.4.1 Morphology and morphometry

Adult had dark brown body [Plate 4.1(A)]. Forewing was slightly convex hard elytra with series of longitudinal pits and hind wing was membranous for flying. Pronotum consisted of posterior grooves giving it a rough appearance and thirteen spines arranged in two ridges anteriorly along the curvature of pronotum like a crown. Pronotum was hood like and covered both prothorax and the entire downward facing head, blocking the view of head from dorsal side. Pronotum and elytra were separated by less sclerotised narrow region. A pair of straight, clubbed antennae consisting of 10 segments with terminal 3 segments forming a club was present in adult. A 'Y'-shaped mark was observed on ventral portion of head. Eyes were black in colour, hemispherical in shape and were bulged out at side of head. Ventrally, five visible abdominal segments

**Table 4.1: Morphometry of lesser grain borer, *R. dominica*** (n=20)

| Stages | | Length (mm) | | | Breadth (mm) | | |
|---|---|---|---|---|---|---|---|
| | | Minimum | Maximum | Mean ± SD | Minimum | Maximum | Mean ± SD |
| Egg | | 0.46 | 0.55 | 0.51 ± 0.04 | 0.18 | 0.23 | 0.19 ± 0.01 |
| Larva | I instar | 0.68 | 0.75 | 0.71 ± 0.01 | 0.15 | 0.19 | 0.18 ± 0.00 |
| | II instar | 1.10 | 1.40 | 1.22 ± 0.06 | 0.36 | 0.47 | 0.40 ± 0.02 |
| | III instar | 1.98 | 2.34 | 2.18 ± 0.06 | 0.56 | 0.75 | 0.63 ± 0.04 |
| | IV instar | 2.58 | 2.72 | 2.68 ± 0.02 | 0.89 | 1.03 | 0.98 ± 0.02 |
| Prepupa | | 2.56 | 2.70 | 2.66 ± 0.02 | 0.88 | 1.01 | 0.97± 0.01 |
| Pupa | | 2.52 | 2.61 | 2.52 ± 0.02 | 0.94 | 1.00 | 0.96 ± 0.01 |
| Adult | Body | 2.36 | 2.54 | 2.45 ± 0.03 | 0.84 | 0.90 | 0.88 ± 0.01 |
| | Pronotum | 0.79 | 0.83 | 0.80 ± 0.03 | 0.75 | 0.79 | 0.77 ± 0.01 |
| | Thorax | 0.96 | 0.99 | 0.98 ± 0.02 | 0.80 | 0.83 | 0.82 ± 0.01 |
| | Elytra | 1.75 | 1.78 | 1.77 ± 0.02 | 0.83 | 0.88 | 0.86 ± 0.01 |
| | Antenna | 0.43 | 0.49 | 0.47 ± 0.03 | - | - | - |
| | Eyes (radius) | 0.10 | 0.12 | 0.10 ± 0.01 | - | - | - |
| | Head capsule | 0.51 | 0.53 | 0.52 ± 0.02 | 0.38 | 0.44 | 0.41 ± 0.01 |
| | Abdomen | 0.85 | 0.90 | 0.88 ± 0.03 | 0.71 | 0.76 | 0.73 ± 0.01 |
| | Average length of each abdominal segment visible | 0.15 | 0.19 | 0.17 ± 0.03 | 0.64 | 0.69 | 0.66 ± 0.01 |

were observed. No difference in morphology among fresh and few days old adults were observed except for pigmentation and sclerotization.

The length and breadth of adult was 2.45 ± 0.03 and 0.88 ± 0.01 mm, respectively. The length and breadth of pronotum was found 0.80 and 0.77 mm, respectively. The length and breadth (both wings combined) of elytra was found 1.77 and 0.86 mm, respectively. The length of metathorax was found 0.58 mm whereas width of abdominal segments ranged between 0.15 to 0.25 mm with total abdominal length of 0.88 mm. Length and breadth of head capsule was 0.52 and 0.41 mm, respectively. The radius of eye was 0.1 mm and length of antennae was 0.47 mm. These findings contribute to a better understanding of the morphological characteristics of this pest and support the existing knowledge in this field and adds valuable insights to the understanding of the lesser grain borer's physical attributes. Similar results on morphology and morphometry were obtained by Edde and Phillips (2006), Chintala *et al.* (2017) and Seada and Hamza (2023) who reported that adults were dark reddish brown in colour and 3 mm long having antennae with three segmented club.

#### 4.1.4.2 Adult longevity

The data on adult longevity presented on Table 4.2 indicated that average life span of adult female and male was 31.85 ± 3.16 and 28.23 ± 2.14 days, respectively. The present investigation was fairly supported by previous works of Kumawat (2007) and Chintala and Virani (2018) who reported that the average life span of female and male was 32.75 ± 2.82, 25.85 ± 2.12 days, respectively.

#### 4.1.4.3 Pre-oviposition period

The minimum and maximum pre-oviposition period recorded was 2 and 4 days, respectively, with average value of 2.5 ± 0.60 days. Data on pre-oviposition period by Chintala *et al.* (2017) indicated average period of 1.75 ± 0.72 days which is more or less similar to the results of present studies.

#### 4.1.4.4Oviposition period

The minimum and maximum oviposition period recorded was 20 and 25 days, respectively, with average value of 22.50 ± 2.23 days. Data on oviposition period by

Chintala *et al.* (2017) indicated average period of 22.05 ± 2.12 days which is similar to the results of present studies.

### 4.1.4.5Post-oviposition period

The minimum and maximum oviposition period recorded was 5 and 9 days, respectively, with average value of 6.85 ± 1.32 days. Data on oviposition period by Chintala *et al.* (2017) indicated average period of 7.95 ± 1.41 days which is more or less similar to the results of present studies.

### 4.1.4.4 Fecundity

The least fecundity observed was 120 eggs per female and highest fecundity observed was 250 eggs per female with average of 191 ± 40.5 eggs per female. The present findings are in close agreement with Kumawat (2007) who revealed that fecundity ranged from 60.0 to 307.3 eggs per female and Win and Rolania (2020) who observed thatfecundity ranged from 123 to 215 (170.8 ± 16.08) eggs in winter and 285 to 398 (306.8 ± 16.36) eggs in the rainy season.

### 4.1.4.5Sex ratio

Male and female adults were sexed based on presence of pale-yellow colour at last few abdominal segments of female [Plate 4.1(C)]. The sex ratio (male : female) was found as 1 :1. The present investigation was supported by Chintala *et al.* (2017) who reported male to female sex ratio of 1.09 : 1.0 whereas, slightly different sex ratio (0.90 : 1.10 ) was reported by Win and Rolania (2020) and it could be due to differences in rearing temperature and lack of easy and reliable sexing methods for lesser grain borer.

### 4.1.4.6Total life cycle

Total life cycle of lesser grain borer is presented in Table 4.2 and Plate 4.2. The data showed that it took an average of 63.55 ± 4.88 and 59.93 ± 3.86 days to complete the life cycle by for female and male, respectively. The data from present investigation is more or less similar to previous studies by Ajaykumara *et al.* (2018) who reported total mean life cycle of 60.96 days.

**Table 4.2: Duration of various life stages of lesser grain borer, *R. dominica* under laboratory conditions**

| Stages | Sample size | Duration (days) | | |
|---|---|---|---|---|
| | | Minimum | Maximum | Mean ± SD |
| Egg | 25 | 5 | 6 | 5.40 ± 0.50 |
| Larva | | | | |
| I instar | 20 | 4 | 6 | 4.30 ± 0.72 |
| II instar | 20 | 3 | 4 | 3.50 ± 0.31 |
| III instar | 20 | 4 | 5 | 4.60 ± 0.56 |
| IV instar | 20 | 4 | 8 | 6.80 ± 0.92 |
| Total larval period | | 20 | 29 | 24.35 ± 2.83 |
| Prepupa | 20 | 1 | 1 | 1.00 ± 0.00 |
| Pupa | 20 | 3 | 6 | 3.85 ± 0.87 |
| Adult | | | | |
| Female | 20 | 27 | 38 | 31.85 ± 3.16 |
| Male | 20 | 26 | 35 | 28.23 ± 2.14 |
| Pre-oviposition period | 20 | 2 | 4 | 2.50 ± 0.60 |
| Oviposition period | 20 | 20 | 25 | 22.50 ± 2.23 |
| Post-oviposition period | 20 | 5 | 9 | 6.85 ± 1.32 |
| Total life cycle | | | | |
| Female | 20 | 50 | 70 | 63.55 ± 4.88 |
| Male | 20 | 49 | 67 | 59.93 ± 3.86 |
| Fecundity | 20 | 120 | 250 | 191 ± 40.5 |
| Sex ratio (male :female) | 20 | 1 :1 | | |
| Hatching percentage | 25 | 80 | | |

## 4.2 SCREENING OF DIFFERENT VARIETIES OF WHEAT FOR SUSCEPTIBILITY TO LESSER GRAIN BORER, *R. dominica*

Ten different varieties of wheat were evaluated for their susceptibility to *R. dominica* based on population growth, weight loss, germination loss, morphological and physical characteristics of wheat up to 180 days of storage.

### 4.2.1 Evaluation Based on Population Growth of Lesser Grain Borer, *R. dominica*

The data obtained from monthly observations on the population buildup of *R. dominica* over a six-month period is presented in Table 4.3.

**Periodical evaluation**

The data presented in Table 4.3 indicated that after 30 days of storage, the variety GW513 exhibited a significantly lower number of emerged adults (12.53)followed by HI 1544 (16.65), which were at par with each other. The next effective order of varieties was LOK 1 (24.80), GW 499 (25.20), GW 451 (29.16), GDW 1255 (29.92), GADW 3 (40.20)GW 206 (43.30) and GW 1 (52.71). The variety GW 496(57.76) showed the highest adult emergence.

After 60 days of storage, the variety GW 513 exhibited the significant lowest number of emerged adults (23.04), followed by HI 1544 (37.70) and GW 499 (34.22). Additionally, the order of adult emergence increased in the following manner without significant differences with each other: GDW 1255 (53.73), GW 451 (57.15), LOK 1 (57.15),GADW 3 (59.75),GW 496 (75.34) and GW 206 (69.39). The highest adult emergence was observed in variety GW 1 (77.44).

The variety, GW 513exhibited the lowest number of emerged adults (31.81) after 90 days of storage followed by varieties GW 499 (45.97) and HI 1544 (51.84) which were at each other. The next effective varieties were GADW 3(65.12), GDW 1255(67.40), GW 1(90.25), GW 496(93.70), GW 206(95.84) and LOK 1(107.95). The significant highest adult emergence of 188.79was observed in the variety GW 451.

After 120 days of storage, the variety GW 513 exhibited significantly lower number of emerged adults (61.31) followed by HI 1544(107.95). GW 499(153.76),

LOK 1(158.51), GW 496(165.38), GDW 1255(165.89) and GADW 3(181.98) were found next effective varieties. GW 451 showed significant highest adult emergence of 316.48.

**Table 4.3: Susceptibility of wheat varieties to *R. dominica* based on periodical population growth of *R. dominica***

| Tr. No. | Varieties | Number of adults emerged after indicated days of storage | | | | | |
|---|---|---|---|---|---|---|---|
| | | 30 | 60 | 90 | 120 | 150 | 180 |
| T1 | GW 513 | $3.54^{e}$ (12.53) | $4.80^{d}$ (23.04) | $5.64^{e}$ (31.81) | $7.83^{f}$ (61.31) | $10.44^{e}$ (108.99) | $13.25^{f}$ (175.56) |
| T2 | GDW 1255 | $5.47^{d}$ (29.92) | $7.33^{b}$ (53.73) | $8.21^{c}$ (67.40) | $12.88^{bc}$ (165.89) | $18.70^{bc}$ (349.69) | $24.85^{ab}$ (617.52) |
| T3 | GADW 3 | $6.34^{c}$ (40.20) | $7.73^{b}$ (59.75) | $8.07^{c}$ (65.12) | $13.49^{b}$ (181.98) | $20.23^{ab}$ (409.25) | $27.36^{a}$ (748.57) |
| T4 | GW 451 | $5.40^{d}$ (29.16) | $7.56^{b}$ (57.15) | $13.74^{a}$ (188.79) | $17.79^{a}$ (316.48) | $21.09^{a}$ (444.79) | $23.90^{bc}$ (571.21) |
| T5 | GW 206 | $6.58^{bc}$ (43.30) | $8.33^{a}$ (69.39) | $9.79^{b}$ (95.84) | $11.93^{cd}$ (142.32) | $14.37^{d}$ (206.50) | $17.03^{d}$ (290.02) |
| T6 | HI 1544 | $4.08^{e}$ (16.65) | $6.14^{c}$ (37.70) | $7.20^{cd}$ (51.84) | $10.39^{e}$ (107.95) | $12.78^{d}$ (163.33) | $14.10^{ef}$ (198.81) |
| T7 | GW 499 | $5.02^{d}$ (25.20) | $5.85^{c}$ (34.22) | $6.78^{d}$ (45.97) | $12.40^{c}$ (153.76) | $18.88^{abc}$ (356.45) | $25.62^{ab}$ (656.38) |
| T8 | GW 1 | $7.26^{ab}$ (52.71) | $8.80^{a}$ (74.44) | $9.50^{b}$ (90.25) | $11.37^{d}$ (129.28) | $13.68^{d}$ (187.14) | $16.65^{d}$ (277.22) |
| T9 | GW 496 | $7.60^{a}$ (57.76) | $8.68^{a}$ (75.34) | $9.68^{b}$ (93.70) | $12.86^{bc}$ (165.38) | $16.89^{c}$ (285.27) | $21.44^{c}$ (459.67) |
| T10 | LOK 1 | $4.98^{d}$ (24.80) | $7.56^{b}$ (57.15) | $10.39^{b}$ (107.95) | $12.59^{bc}$ (158.51) | $14.45^{d}$ (208.80) | $16.12^{de}$ (259.85) |
| S. Em. ± | | 0.24 | 0.17 | 0.36 | 0.28 | 0.67 | 0.76 |
| F Test (T) | | Sig. | Sig. | Sig. | Sig. | Sig. | Sig. |
| C.V. (%) | | 7.30 | 3.96 | 7.10 | 3.97 | 7.14 | 6.55 |

**Notes:** 1. Figures in parentheses are retransformed values of square root transformed values
2. Treatment mean(s) with letter(s) in common are not differed significantly by Duncan's New Multiple Range Test (DNMRT) at 5% level of significance

After 150 days of storage, significantly lower adult emergence of 108.99 was recorded in the variety GW 513. Further, adult emergence increased without significant differences in the order: HI 1544 (163.33), GW 1 (187.14), GW 206 (206.50),LOK 1 (208.80). GW 496 and GDW 1255 were at par with each other for adult emergence of 285.27 and 349.69, respectively. GW 451 showed the significant highest adult emergence of 444.79 which was at par with GADW 3 (409.25)and GW 499 (356.45).

Significantly lower population growth was observed in varieties GW 513 and HI 1544 after 180 days of storage with adult emergence of 175.56 and 198.81, respectively. Further, increase in population growth without significant differences was exhibited by the varieties LOK 1 (259.85), GW 1 (277.22) and GW 206 (290.02) followed by GW 496 (459.67) and GW 451 (571.21). The varieties GDW 1255, GW 499 and GADW 3were found with significantly higher population growth with adult emergence of 617.52, 656.38 and 748.57, respectively.

The current investigation on population growth of *R. dominica* on stored wheat found relative susceptibility among the varieties evaluated in previous works, in the order: GW 499 > LOK 1 > HI 1544 which was supported by research work carried out by Chintala (2015), Kher (2006) and Baghla *et al.* (2023). Scanty information is available regarding population buildup of *R. dominica* on other wheat varieties under investigation. The observed higher number of adults in the present investigation compared to previous studies conducted by Kher (2006), Chintala (2015) and Baghla *et al.* (2023) could be attributed to the potential differences in the reproductive capacity of the pest population used in the experiment.

### 4.2.2 Evaluation Based on Weight Loss of Grains

The data obtained from investigation on per cent weight loss of wheat grains due to *R. dominica* after six months of storage period is presented in Table 4.4. The variety GW 513 recorded significant the lowest percentage of weight loss, measuring 10.72%, followed by next effective variety 1544 (12.97%). The variety LOK 1 shown a weight loss of 15.25 % and was followed by GW 206 (16.18) and GW 1 (16.03). The varieties, GW 496 and 451 GDW 1255 were recorded per cent weight loss of 19.98, 21.94 and 22.26, respectively and were at par with each other followed by GW 499 (26.23). GADW 3 exhibited the highest weight loss of 35.31 per cent.

**Table 4.4:Susceptibility of wheat varieties to *R. dominica* based on weight loss**

| Tr. No. | Varieties | Seed weight (g) | | Weight loss (%) |
|---|---|---|---|---|
| | | Before adult release | After six months of storage under infestation | |
| **T1** | **GW513** | 4.79 | 4.27 | 19.11[f] (10.72) |
| **T2** | **GDW 1255** | 5.51 | 4.25 | 28.15[c] (22.26) |
| **T3** | **GADW3** | 5.86 | 3.77 | 36.46[a] (35.31) |
| **T4** | **GW 451** | 4.62 | 3.62 | 27.93[c] (21.94) |
| **T5** | **GW206** | 4.11 | 3.42 | 23.72[d] (16.18) |
| **T6** | **HI1544** | 4.32 | 3.62 | 21.11[e] (12.97) |
| **T7** | **GW499** | 5.14 | 3.79 | 30.81[b] (26.23) |
| **T8** | **GW1** | 4.55 | 3.82 | 23.60[d] (16.03) |
| **T9** | **GW496** | 5.00 | 4.00 | 26.55[c] (19.98) |
| **T10** | **LOK1** | 4.91 | 4.16 | 22.99[de] (15.25) |
| **S. Em. ±** | | 0.14 | 0.14 | 0.67 |
| **F Test (T)** | | Sig. | Sig. | Sig. |
| **C.V. (%)** | | 5.06 | 6.15 | 4.46 |

**Notes:** 1. Figures in parentheses are retransformed values of arc sine transformed values

2. Treatment mean(s) with letter(s) in common are not differed significantly by Duncan's New Multiple Range Test (DNMRT) at 5% level of significance

Earlier, very little work has been done on evaluation of susceptibility of the wheat varieties used in present study. However, weight loss percentage of LOK 1,GW 496 and GADW 3 from present investigation is in accordance with studies by Kher (2006), Baghla *et al.* (2023) and Gami *et al.* (2022), respectively.

### 4.2.3 Evaluation Based on Reduction in Seed Germination

The data on percent germination of various wheat varieties before and after artificial infestation along with reduction in germination following infestation by *R. dominica* after six months of storage is presented in Table 4.5.

The data on germination percentages before artificial infestation revealed a range of 87.67 % to 94.44%, with an average germination rate of 90.75 %. The varieties did not differ significantly from each other in terms of germination percentages before artificial infestation.

The data on percent germination after six months of storage with infestation by *R. dominica* was found significant among the varieties and it ranged from 43.3 % to 12.26 % with an average of 25.32 %. The highest germination percentage was recorded in the variety GW 513 (43.3) and was at par with GW 206 (42.30). GDW 1255 resulted in 28.96 % germination and was at par with GW 496 (27.64 %) and GW 1 (26.31 %) followed by GW 499 (25.27 %) and GADW 3 (22.6 %). LOK 1 resulted in germination percentage of 18.63 followed by GW 451 (12.26) and least per cent germination was reported in the variety HI 1544 (5.99).

The per cent germination loss due to infestation by*R. dominica* after six months of release revealed that the significantly lower per cent reduction in germination (50.54)was observed in GW 513 which was at par with GW 206 (54.71). There was no significant difference in per cent loss of germination among the varieties GDW 1255 (68.94), GW 1 (69.89), GW 496 (69.95) GADW 3 (73.55) and GW 499 (72.65). Variety, LOK 1 (78.66) was followed by GW 451 (86.50). The highest germination loss (93.30 %) was reported in the variety HI 1544.

Similar data on germination loss in case of the varieties LOK 1 and GW 499 were reported by Kher (2006), however GADW 3was varied in this experiment as reported by Gami *et al.* (2022), which could be due to change in location and environment.

**Table 4.5:Susceptibility of wheat varieties to *R. dominica* based on germination**

| Tr. No. | Varieties | Germination (%) | | Germination loss (%) |
|---|---|---|---|---|
| | | Before adult release | After six months of storage under infestation | |
| T1 | GW 513 | 69.47 (87.7) | 41.15[a] (43.3) | 45.31[e] (50.54) |
| T2 | GDW 1255 | 75.96 (94.11) | 32.56[b] (28.96) | 56.13[d] (68.94) |
| T3 | GADW 3 | 69.44 (87.67) | 28.63[d] (22.6) | 59.05[d] (73.55) |
| T4 | GW 451 | 73.22 (91.67) | 20.50[f] (12.26) | 68.44[b] (86.50) |
| T5 | GW 206 | 76.36 (94.44) | 40.57[a] (42.30) | 47.70[e] (54.71) |
| T6 | HI 1544 | 71.55 (89.98) | 14.17[g] (5.99) | 75.00[a] (93.30) |
| T7 | GW 499 | 75.16 (93.44) | 30.18[cd] (25.27) | 58.47[d] (72.65) |
| T8 | GW 1 | 69.51 (87.75) | 30.86[bc] (26.31) | 56.72[d] (69.89) |
| T9 | GW 496 | 74.35 (92.72) | 31.72[bc] (27.64) | 56.76[d] (69.95) |
| T10 | LOK 1 | 69.57 (87.82) | 25.57[e] (18.63) | 62.49[c] (78.66) |
| S. Em. ± | | 2.53 | 0.65 | 2.68 |
| F Test (T) | | NS | Sig. | Sig. |
| C.V. (%) | | 6.07 | 3.81 | 2.68 |

**Notes:** 1. Figures in parentheses are retransformed values of arc sine transformed values

2. Treatment mean(s) with letter(s) in common are not differed significantly by Duncan's New Multiple Range Test (DNMRT) at 5% level of significance

**Table 4.6:Effect of morphological characters of different varieties on weight loss due to *R. dominica* in wheat**

| Sr. No. | Varieties | Length (mm) | Width (mm) | 100 Seed Weight (g) | Seed Hardness (N) | Weight loss (%) |
|---|---|---|---|---|---|---|
| 1 | **GW 513** | 6.92 | 3.39 | 12.31 | 20.92 | 13.64 |
| 2 | **GDW 1255** | 7.55 | 3.50 | 13.57 | 12.68 | 12.68 |
| 3 | **GADW 3** | 9.55 | 3.26 | 14.01 | 15.23 | 15.23 |
| 4 | **GW 451** | 6.81 | 3.55 | 12.80 | 12.67 | 15.65 |
| 5 | **GW 206** | 8.05 | 3.33 | 12.92 | 13.76 | 12.04 |
| 6 | **HI 1544** | 6.43 | 3.27 | 11.69 | 17.25 | 13.76 |
| 7 | **GW 499** | 7.44 | 3.37 | 13.10 | 12.36 | 12.36 |
| 8 | **GW 1** | 7.84 | 3.23 | 11.99 | 13.64 | 17.25 |
| 9 | **GW 496** | 6.60 | 3.52 | 12.64 | 12.04 | 20.92 |
| 10 | **LOK 1** | 7.92 | 3.47 | 12.41 | 15.65 | 12.67 |
| **S. Em. ±** | | 0.28 | 0.13 | 0.21 | 0.56 | 0.92 |
| **F Test (T)** | | Sig. | NS | Sig. | Sig. | Sig. |
| **C.V. (%)** | | 6.42 | 6.72 | 7.30 | 6.69 | 8.05 |
| **Correlation coefficient (r)** | | 0.73* | 0.61 | 0.87** | -0.48 | - |

**Notes:** 1) N: Newton

2) * Significant at 5 % level of significance

3) **Significant at 1 % level of significance

**Based on population growth**

Mean value of population growth of all the varieties was 11.73 and 2.18, respectively. None of the varieties were found either highly resistant or highly susceptible as population growth was within the range of 7.36 and 36.23. Population growth of GW 513 and HI 1544 was greater than 7.36 and less than or equal to 9.54 and were considered as resistant. The varieties GW 206, GW 1 and LOK 1 were considered moderately resistant as the mean population growth was greater than 9.54 and less than or equal to 11.72. The moderately susceptible varieties were GDW 1255, GADW 3, GW 499 and GW 496 which had adult emergence greater than 11.72 and less than or equal to 13.91. The variety, GW 451 was found susceptible.

**Based on weight loss**

Mean value of population growth of all the varieties was 26.06 and 5.08, respectively. None of the variety was found highly resistant as all the varieties exhibited weight loss of value greater than 15.89. The variety GW 513 was found resistant as weight loss was neither less than or equal to 15.89 nor greater than 20.97. The varieties, GW 206, HI 1544, GW 1 and LOK 1were categorized as moderately resistant because their per cent weight loss was greater than 20.97 and less than or equal to 26.06. The adult emergence in the varieties GW 451, GW 496, GDW 1255 and GW 499 was greater than 26.06 and less than or equal to 31.14 and therefore they were grouped under moderately susceptible category. None of the variety was categorized as susceptible, however, the variety GADW 3 showing per cent weight loss greater than 36.23 was categorized as highly susceptible.

**Based on germination loss**

Mean value of per cent germination loss of all the varieties was 58.64 and 8.75, respectively. All the varieties exhibited per cent germination loss above 41.12, hence no variety was found highly resistant based on germination loss. The variety GW 513 and GW 206 were found resistant as per cent germination loss was neither less than or equal to 41.01 nor greater than 49.73. The varieties, GDW 1255, GADW 3, GW 1 and GW 496 were categorized as moderately resistant because their per cent germination loss was greater than 49.73 and less than or equal to 58.44. The adult emergence in the varieties, GW 499 and LOK 1was greater than 58.44 and less than or equal to 67.15and hence are grouped under moderately susceptible category.

**Table 4.7: Categorization of different varieties of wheat for their susceptibility againstlesser grainborer**

| Categories of resistance | Scale | Varieties |
|---|---|---|
| **Based on population growth** | | |
| | $\bar{x} = 11.73$ SD = 2.18 | |
| **Highly resistant** | $\overline{xi} \leq 7.36$ | Nil |
| **Resistant** | $7.36 < \overline{xi} \leq 9.54$ | GW 513 and HI 1544 |
| **Moderately resistant** | $9.54 < \overline{xi} \leq 11.72$ | GW 206, GW 1, LOK 1 |
| **Moderately susceptible** | $11.72 < \overline{xi} \leq 13.91$ | GDW 1255, GADW 3, GW 499, GW 496 |
| **Susceptible** | $13.91 < \overline{xi} \leq 16.09$ | GW 451 |
| **Highly susceptible** | $\overline{xi} > 16.09$ | Nil |
| **Based on weight loss** | | |
| | $\bar{x} = 26.06$SD = 5.08 | |
| **Highly resistant** | $\overline{xi} \leq 15.89$ | Nil |
| **Resistant** | $15.89 < \overline{xi} \leq 20.97$ | GW 513 |
| **Moderately resistant** | $20.97 < \overline{xi} \leq 26.06$ | GW 206, HI 1544, GW 1, LOK 1 |
| **Moderately susceptible** | $26.06 < \overline{xi} \leq 31.14$ | GW 451, GW 496, GDW 1255, GW 499, |
| **Susceptible** | $31.14 < \overline{xi} \leq 36.23$ | Nil |
| **Highly susceptible** | $\overline{xi} > 36.23$ | GADW 3 |
| **Based on germination loss** | | |
| | $\bar{x} = 58.64$SD = 8.75 | |
| **Highly resistant** | $\overline{xi} \leq 41.12$ | Nil |
| **Resistant** | $41.12 < \overline{xi} \leq 49.88$ | GW 513, GW 206 |
| **Moderately resistant** | $49.88 < \overline{xi} \leq 58.64$ | GDW 1255, GADW 3, GW 1, GW 496 |
| **Moderately susceptible** | $58.64 < \overline{xi} \leq 67.40$ | GW 499, LOK 1 |
| **Susceptible** | $67.40 < \overline{xi} \leq 76.16$ | GW 451, HI 1544 |
| **Highly susceptible** | $\overline{xi} > 76.16$ | Nil |

The varieties, GW 451 and HI 1544 were grouped under susceptible category as per cent germination loss was found greater than 31.14 and less than or equal to 75.86, however, none of the varieties were categorized as highly susceptible based on per cent germination loss as it did not exceed 75.86.

The variety, GW 513 was found resistant based on germination loss and population growth, whereas it was moderately resistant based on weight loss. The variety, GW 1 was found moderately resistant based on all three parameters. The variety, GDW 1255 was moderately susceptible based on population growth and weight loss, whereas, it was moderately resistant based on germination loss. The variety, LOK 1 was found moderately resistant based on population growth and weight loss, whereas it was moderately susceptible based on germination loss. The variety, GW 451 was found susceptible based on population growth and germination loss, whereas it was moderately resistant based on weight loss. Both varieties, GW 496 and GADW 3 were found moderately susceptible based on population growth, whereas it was moderately resistant based on germination loss. The variety, HI 1544 was found resistant based on population growth, moderately resistant based on weight loss and susceptible based on germination loss. The variety GW 206 was found moderately resistant based on population growth and weight loss, whereas it was resistant based on germination loss. None of the varieties under investigation was found highly resistant based on the three parameters considered for evaluation and categorization of resistance, however the variety GADW 3 was categorized as high susceptibility based on weight loss.

The category of resistance could not be exactly compared with previous works because only a few varieties from present investigation have been studied previously for their susceptibility to lesser grain borer and relative susceptibility varies with the varieties under evaluation. However, the order of susceptibility or relative resistance of identical varieties can be compared. Higher relative susceptibility of LOK 1 based on weight loss and germination loss in comparison with GW 496 was reported by Kher (2006). Whereas, higher relative susceptibility of GW 496 based on population growth in comparison with LOK 1 was reported by Kher (2006) and Chintala (2015). Baghla *et al.* (2023) reported greater resistance of HI 1544 based on weight loss and population growth in comparison with LOK 1. Hence, results of the present findings are in accordance with the work done by earlier researchers.

## 4.3 EVALUATION OF DIFFERENT NON-EDIBLE OILS AND ITS NANO-EMULSIONS AS GRAIN PROTECTANTS AGAINST LESSER GRAIN BORER, *R. dominica* ON STORED WHEAT

Four non-edible oils and their nano-emulsions(Table 3.3) at 1 per cent (v/w) level were tested as grain protectants against *R. dominica*on wheat under storage conditions in laboratory. The efficacy of grain protectants based on their effect on adult mortality and population growth was carried out.

### 4.3.1Preparationand Characterization of Nano-Emulsions

Nano-emulsions were prepared according to the methodology mentioned in 3.3.3. The properties of obtained nano-emulsions are presented in Table 4.8 and Figures 4.1 to 4.8. All the obtained nano-emulsions were milky or opaque white in colour and did not flocculate when stored at 6 % concentration at 5 °C in a refrigerator for a period of six months.

The physical properties of the nano-emulsions prepared from castor oil, neem oil, karanj oil and eucalyptus oil were evaluated. The hydrodynamic size of the nano-emulsions was found to be 64.07 ± 1.32 nm, 37.04 ± 0.75 nm, 33.90 ± 0.29 nm and 46.56 ± 1.25 nm, respectively. The zeta potential of the nano-emulsions was -41.00± 6.45 mV, -53.90± 6.30 mV, -54.90± 6.25 mV, and -30.90 ± 6.82 mV for castor oil, neem oil, karanj oiland eucalyptus oil, respectively. The kilo count per second was measured as 291.30 ± 1.20, 297.80 ± 1.40, 271.70 ± 0.30and 305.60 ± 3.10 for castor oil, neem oil, karanj oiland eucalyptus oil, respectively. The polydispersity index values were determined as 0.318 ± 0.07, 0.397 ± 0.02, 0.345 ± 0.01and 0.390 ± 0.02 for castor oil, neem oil, karanj oiland eucalyptus oil, respectively.

These results indicated that hydrodynamic size of all the prepared nano-emulsions found within the range of nano-scale (1 – 999 nm).Further more,zeta potential was above standard value of ±30 mV for all the nano-emulsions and hence, it indicated particle stability and repulsion, preventing flocculation. Higher kilo count per second values generally correspond to a higher concentration of dispersed particles in the nano-emulsion and the kilo count per second values from present investigation suggested that eucalyptus oil nano-emulsion had the highest particle concentration, followed by neem oil, castor oil, and karanj oil nano-emulsions.The polydispersity index value was less than the standard limit of 0.5 and it suggested less variation in particle size, confirming the stability of the prepared nano-emulsions.

Similar values of polydispersity index were reported byAdak *et al.* (2020) and Iqbal *et al.* (2022) in case of eucalyptus nano-emulsion and neem nano-biopesticide which resulted in mortality of *S. oryzae* and *R. dominica,* respectively. Similar zeta potential values were obtained by Iqbal *et al.* (2022) whereas Adak *et al.* (2020) recorded lower and positive zeta potential and it could be due to difference in ratio between oil and surfactant and method of particle size reduction.

**Table 4.8: Physical characterization of nano-emulsions**

| Nano emulsions | Hydrodynamic size (nm) | Zeta potential (mV) | Kilo count per second | Polydispersity index |
|---|---|---|---|---|
| Castor oil nano-emulsion (6 %) | 64.07 ± 1.32 | -41.00± 6.45 | 291.30 ± 1.20 | 0.318 ± 0.07 |
| Neem oil nano-emulsion (6 %) | 37.04 ± 0.75 | -53.90± 6.30 | 297.80 ± 1.40 | 0.397 ± 0.02 |
| Karanj oil nano-emulsion (6 %) | 33.90 ± 0.29 | -54.90± 6.25 | 271.70 ± 0.30 | 0.345 ± 0.01 |
| Eucalyptus oil nano-emulsion (6 %) | 46.56 ± 1.25 | -30.90 ± 6.82 | 305.60 ± 3.10 | 0.390 ± 0.02 |

### 4.3.2 Evaluation of Non-Edible Oils and Nano-Emulsions Based on Adult Mortality of *R. dominica*

The data obtained during experiment on per cent corrected mortality of *R. dominica* carried out at 15 days interval for six month storage duration is presented in Table 4.9.

**Periodical evaluation**

The data presented in Table 4.10 indicated that the treatments differed significantly from each other after 15 days of storage. Neem oil nano-emulsion exhibited significantly higher adult mortality (25.62 %) followed by karanj oil nano-emulsion (21.75 %). Castor oil nano-emulsion and neem oil were at par with each other with corrected mortality per cent of 16.49 and 15.37, respectively. Karanj oil exhibited corrected mortality per cent of 13.15 which was followed by eucalyptus oil nano-emulsion (8.10) and castor oil (8.02). Lower mortality among oils and nano-emulsions was observed from the treatment eucalyptus oil (6.32 %), whereas, lowest corrected mortality per cent of 0.01 was recorded from the untreated seed treatment.

After 30 days of storage, neem oil nano-emulsion exhibited significantly higher adult mortality (31.14 %)followed by castor oil nano-emulsion, karanj oil nano-emulsion, neem oil and karanj oil with corrected mortality per cent of 23.21, 21.48, 21.11 and 18.89, respectively. Castor oil and eucalyptus oil nano-emulsion exhibited corrected mortality per cent of 13.5 and 13.11,respectively and remained at par with each other. Lower mortality among oils and nano-emulsions was observed from the treatment eucalyptus oil (8.23%), whereas the lowest corrected mortality per cent of 0.01 was recorded from the control treatment.

After 45 days of storage, karanj oil nano-emulsion exhibited highest adult mortality(41.59 %) which was at par with neem oil nano-emulsion (44.18 %). Castor oil nano-emulsion and neem oil were remained at par with each other and exhibited corrected mortality per cent of 28.35and25.91,respectively.Karanj oil and castor oil were found next effective treatments which were at par with each other and exhibited corrected mortality per cent of 21.16 and 16.32,respectively.The lower mortality among oils and nano-emulsions was observed from the treatment eucalyptus oil (11.73 %) which was at par with eucalyptus oil nano-emulsion(15.46 %), whereas the lowest corrected mortality per cent of 0.01 was recorded from the treatment control.

The highest per cent mortality (50.24 %) was recorded from the treatment neem oil nano-emulsion which was at par with karanj oil nano-emulsion(46.32 %) after 60 days of storage. Castor oil nano-emulsion resulted in 34.07 per cent mortality which was followed by neem oil (27.88 %) and karanj oil (22.86 %).Eucalyptus oil nano-emulsion (18.59 %) was found at par with castor oil (18.22 %). The lower mortality among oils and nano-emulsions was observed from the treatment eucalyptus oil (13.46 %), whereas the lowest corrected mortality per cent of 0.01 was recorded from the treatment control.

After 75 days of storage, neem oil nano-emulsion exhibited the highest adult mortality(53.30 %) which was at par with karanj oil nano-emulsion (48.29 %). Castor oil nano-emulsion (36.62 %) and neem oil (31.61 %) were remained at par with each other. Karanj oil and eucalyptus oil nano-emulsion exhibited per cent mortality of

26.60and 21.59, respectively. The lower mortality among oils and nano-emulsions was observed from the treatments, castor oil (19.81 %) and eucalyptus oil (14.94 %), whereas the lowest corrected mortality per cent of 0.01 was recorded from the treatment control.

After 90 days of storage, neem oil nano-emulsion exhibited the highest adult mortality (52.44 %) which was at par with karanj oil nano-emulsion (47.38 %) followed by castor oil nano-emulsion (37.23 %). Neem oil (30.30 %) and karanj oil (25.30 %) were found at par with each other. Eucalyptus oil nano-emulsion and castor oil exhibited per cent mortality of 21.90 and 18.53, respectively. The lower mortality among oils and nano-emulsions was observed from the treatment eucalyptus oil (11.70 %), whereas the lowest corrected mortality per cent of 0.01 was recorded from the treatment control.

After 105 days of storage, the similar trend was observed as per earlier. However, karanj oil, eucalyptus oil nano-emulsion and castor oil exhibited per cent mortality of 23.27, 22.65and18.25,respectively which remained at par with each other. Lower mortality among oils and nano-emulsions was observed from the treatment as per earlier.

After 120 and 135 days of storage, again the similar trend was observed in which neem oil nano-emulsion exhibited highest adult mortality (52.18 % and 49.08 %) which was at par with karanj oil nano-emulsion (47.02 % and 45.75 %) followed by castor oil nano-emulsion (32.52 % and 32.00 %). The rest of the treatments also followed similar trend as per the results obtained earlier.

After 150 days of storage, neem oil nano-emulsion exhibited highest adult mortality(49.97 %) which was at par with karanj oil nano-emulsion (44.96 %) followed by castor oil nano-emulsion (31.61 %). Neem oil, karanj oil and eucalyptus oil nano-emulsion exhibited per cent mortality of 23.27, 19.92and 18.25, respectively which were at par with each other. The lower mortality among oils and nano-emulsions was observed from the treatments castor oil (16.63 %) and eucalyptus oil (8.15%), whereas the lowest corrected mortality per cent of 0.01 was recorded from the treatment control.

After 165 and 180 days of storage, neem oil nano-emulsion exhibited highest adult mortality (50.10 and 52.88 %) followed by karanj oil nano-emulsion (42.04 and 42.40%), castor oil nano-emulsion (29.52 and 31.35 %) and neem oil(21.61 and 21.68

**Table 4.9: Effectiveness ofoils and nano-emulsions as grain protectants based on periodical adult mortality of *R. dominica*during storage of wheat**

| Tr. No. | Treatments | Dose (ml/100 g seeds) | Corrected mortality (%) at indicated days of storage | | | | | |
|---|---|---|---|---|---|---|---|---|
| | | | 15 | 30 | 45 | 60 | 75 | 90 |
| T1 | Castor oil 100 % | 1 | $16.45^{e}$ (8.02) | $21.56^{c}$ (13.50) | $23.83^{de}$ (16.32) | $25.27^{d}$ (18.22) | $26.43^{e}$ (19.81) | $25.50^{e}$ (18.53) |
| T2 | Neem oil 100 % | 1 | $23.08^{cd}$ (15.37) | $27.35^{b}$ (21.11) | $30.60^{bc}$ (25.91) | $31.87^{c}$ (27.88) | $34.21^{bc}$ (31.61) | $33.43^{c}$ (30.30) |
| T3 | Karanj oil 100 % | 1 | $21.26^{d}$ (13.15) | $25.76^{b}$ (18.89) | $27.39^{cd}$ (21.16) | $28.56^{cd}$ (22.86) | $31.05^{cd}$ (26.60) | $30.27^{cd}$ (25.30) |
| T4 | Eucalyptus oil 100 % | 1 | $14.56^{f}$ (6.32) | $16.67^{d}$ (8.23) | $20.03^{f}$ (11.73) | $21.52^{e}$ (13.46) | $22.74^{f}$ (14.94) | $20.04^{f}$ (11.70) |
| T5 | Castor oil nano-emulsion 6 % | 1 | $23.96^{c}$ (16.49) | $28.80^{b}$ (23.21) | $32.17^{b}$ (28.35) | $35.71^{b}$ (34.07) | $37.24^{b}$ (36.62) | $37.60^{b}$ (37.23) |
| T6 | Neem oil nano-emulsion 6 % | 1 | $30.41^{a}$ (25.62) | $33.92^{a}$ (31.14) | $41.66^{a}$ (44.18) | $45.14^{a}$ (50.24) | $46.89^{a}$ (53.30) | $46.40^{a}$ (52.44) |
| T7 | Karanj oil nano-emulsion 6 % | 1 | $27.80^{b}$ (21.75) | $27.61^{b}$ (21.48) | $40.16^{a}$ (41.59) | $42.89^{a}$ (46.32) | $44.02^{a}$ (48.29) | $43.52^{a}$ (47.38) |
| T8 | Eucalyptus oil nano-emulsion 6 % | 1 | $16.53^{e}$ (8.10) | $21.23^{c}$ (13.11) | $23.15^{ef}$ (15.46) | $25.54^{d}$ (18.59) | $27.69^{de}$ (21.59) | $27.94^{de}$ (21.90) |
| T9 | Control | 1 | $0.64^{g}$ (0.01) | $0.64^{e}$ (0.01) | 0.64g (0.01) | $0.64^{f}$ (0.01) | $0.64^{g}$ (0.01) | $0.64^{g}$ (0.01) |
| S. Em. ± | | | 0.55 | 0.95 | 1.10 | 1.03 | 1.15 | 1.20 |
| F Test (T) | | | Sig. | Sig. | Sig. | Sig. | Sig. | Sig. |
| C.V. (%) | | | 4.94 | 7.27 | 7.18 | 6.26 | 6.62 | 7.04 |

**Notes:** 1. Figures in parentheses are retransformed values of arc sine transformed values

2. Treatment mean(s) with letter(s) in common are notdiffered significantly by Duncan's New Multiple Range Test (DNMRT) at 5% level of significance

**Table 4.9: Continued**

| Tr. No. | Treatments | Dose (ml/100 gseeds) | Corrected mortality (%) after indicated days of storage | | | | | | |
|---|---|---|---|---|---|---|---|---|---|
| | | | 105 | 120 | 135 | 150 | 165 | 180 | Pooled |
| T1 | Castor oil 100 % | 1 | 25.29[d] (18.25) | 23.08[d] (15.37) | 22.86[e] (15.09) | 24.07[d] (16.63) | 18.13[fg] (9.68) | 18.34[fg] (9.90) | 22.57[f] (14.73) |
| T2 | Neem oil 100 % | 1 | 33.26[c] (30.08) | 33.39[b] (30.29) | 30.21[c] (25.32) | 28.84[c] (23.27) | 27.70[d] (21.61) | 27.75[d] (21.68) | 30.14[d] (25.10) |
| T3 | Karanj oil 100 % | 1 | 28.84[d] (23.27) | 28.84[c] (23.27) | 26.80[cd] (20.33) | 26.51[cd] (19.92) | 23.52[e] (15.93) | 23.53[e] (15.94) | 26.86[e] (20.41) |
| T4 | Eucalyptus oil 100 % | 1 | 18.23[e] (9.79) | 17.86[e] (9.41) | 16.70[f] (8.26) | 16.59[e] (8.15) | 16.55[g] (8.11) | 16.53[g] (8.10) | 18.17[g] (9.72) |
| T5 | Castor oil nano-emulsion 6 % | 1 | 38.26[b] (38.34) | 34.77[b] (32.52) | 34.45[b] (32.00) | 34.21[b] (31.61) | 32.91[c] (29.52) | 34.05[c] (31.35) | 33.68[c] (30.75) |
| T6 | Neem oil nano-emulsion 6 % | 1 | 46.98[a] (53.45) | 46.25[a] (52.18) | 44.47[a] (49.08) | 44.98[a] (49.97) | 45.06[a] (50.10) | 46.65[a] (52.88) | 43.23[a] (46.91) |
| T7 | Karanj oil nano-emulsion 6 % | 1 | 43.82[a] (47.94) | 43.29[a] (47.02) | 42.56[a] (45.75) | 42.11[a] (44.96) | 40.42[b] (42.04) | 40.63[b] (42.40) | 39.90[b] (41.15) |
| T8 | Eucalyptus oil nano-emulsion 6 % | 1 | 28.42[d] (22.65) | 27.69[c] (21.59) | 25.50[de] (18.53) | 25.29[cd] (18.25) | 21.29[ef] (13.18) | 19.73[f] (11.40) | 24.17[f] (16.76) |
| T9 | Control | 1 | 0.64[f] (0.01) | 0.64[f] (0.01) | 0.64[g] (0.01) | 0.64[f] (0.01) | 0.64[h] (0.01) | 0.64[h] (0.01) | 0.64[h] (0.01) |
| S. Em. ± | Treatments (T) | | 1.19 | 1.25 | 1.04 | 1.07 | 1.11 | 0.95 | 0.36 |
| | | Period (P) | - | - | - | - | - | - | 1.07 |
| | | T × P | - | - | - | - | - | - | 0.67 |
| F Test (T) | | | Sig. | Sig. | Sig. | Sig. | Sig. | Sig. | Sig. |
| C.V. (%) | | | 7.07 | 7.64 | 6.64 | 6.85 | 7.71 | 6.50 | 6.95 |

**Notes:** 1. Figures in parentheses are retransformed values of arc sine transformed values

2. Treatment mean(s) with letter(s) in common are not differed significantly by Duncan's New Multiple Range Test (DNMRT) at 5% level of significance 3. Significant parameters and its interactions, T × P; Sig.: significant

%) which significantly differed from each other. Karanj oil (15.93 and 15.94 %) and eucalyptus oil nano-emulsion (13.18 and 11.40%) remained at par with each other after 165 days, whereas they differed significantly after 180 days. The lower mortality among oils and nano-emulsions was observed from the treatments of castor oil (9.68 and 9.90 %) and eucalyptus oil (8.11 and 8.10%), whereas the lowest corrected mortality per cent (0.01 and 0.01 %) was recorded from the treatment control.

**Pooled over periods**

The periodic data on per cent adult mortality of *R. dominica* were pooled and presented in Table 4.9.The pooled over period data on adult mortality showed that neem oil nano-emulsion (46.91%) was significantly superior over other treatments and found the most effective treatment in terms of adult mortality. Karanjoil nano-emulsion (41.75%), castor oil nano-emulsion (30.75 %), neem oil (25.10 %) and karanj oil(20.41 %)were found next effective than rest of the other treatments. Further, eucalyptus oil nano-emulsion(16.76 %) and castor oil(14.73 %)did not differ significantly with each other andhence proved equally effective. Eucalyptus oil recorded significant lower (9.72) percent adult mortality of *R. dominica* when compared with rest of other oils and nano-emulsions. Whereas, the lowest corrected mortality per cent (0.01) was recorded from the treatment control. Thus, eucalyptus oil was proved as the least effective treatment among oils and nano emulsions under investigation against *R. dominica*in terms of adult mortality.

Overall effectiveness of the treatments under investigation were found in the order: neem oil nano-emulsion>karanjoil nano-emulsion>castor oil nano-emulsion>neem oil>karanj oil> eucalyptus oil nano-emulsion> castor oil> eucalyptus oil> control. Hence, it was found that all the nano-emulsions and oils were effective as grain protectants in terms of adult mortality. All nano-emulsions were found more effective than oils except eucalyptus oil nano-emulsion, however, nano-emulsions of corresponding oils were more effective than the oils, thus proving nano-emulsions as a better alternative to oils as grain protectants.

Very little information is available in literature on comparison of various nano-emulsions as grain protectants of wheat against *R. dominica*, therefore, only oils and their efficacy could be compared with earlier research. The overall order of effectiveness of various plant oils as grain protectants of wheat against *R. dominica* found from current investigation matches with previous works by Kher (2006), Nema and Bhargava (2012), Kumawat and Naga (2013) and Chintala (2015). However, it slightly differed when compared with results obtained by Singh *et al.* (2016) which could be due to change in quality of oils used or the variety of wheat used in the experiment. Data on per cent mortality of eucalyptus oil from present studies was not in accordance with Chandel *et al.* (2021) which could be due to the use of different concentrations of oils for assessment of mortality. Enhanced effectiveness of nano-emulsions in comparison with regular oil as found from current investigation is supported by Sabbour and El-Aziz (2016), Adak *et al.* (2020), Sharma *et al.* (2020) and Iqbal *et al.* (2022).

### 4.3.3 Evaluation of The Non-Edible Oils and Nano-Emulsions Based on Population Growth of *R. dominica*

The data obtained from monthly observations on the population buildup of *R. dominica* over a six-month period are presented in Table 4.10.

**Periodical evaluation**

The data on number of adults developed after 30 days of storage revealed that neem oil nano-emulsion resulted in least number of adults emerged (1.61) which was at par with karanj oil nano-emulsion (1.99). Castor oil nano-emulsion exhibited adult emergence of 2.62 which was followed by eucalyptus oil nano-emulsion (2.99), neem oil (2.99), karanj oil(3.31) and castor oil(3.31). The significant higher number of adults were emerged from control (6.97).

After 60 days of storage, least number of adults (2.62) were emerged from neem oil nano-emulsion followed by karanj oil nano-emulsion (3.65) and both significantly differed from each other. Castor oil nano-emulsion exhibited adult emergence of 4.97 which was at par with eucalyptus oil nano-emulsion (5.95) and neem oil(5.95). The treatments, castor oil(6.30), karanj oil(6.66) and eucalyptus oil (7.29) were at par with each other. The significantly higher number of adults were emerged from control (17.56).

After 90 days of storage, least number of adults (4.97) emerged from neem oil nano-emulsion. Karanj oil nano-emulsion and castor oil nano-emulsion exhibited adult emergence values of 6.97 and7.95, respectively and remained at par with each other. Neem oil(11.29), eucalyptus oil nano-emulsion (11.90), karanj oil (11.97) and castor oil (12.60) were found at par with each other. Eucalyptus oil (14.29) exhibited higher population growth among oils and nano-emulsions, whereas significantly higher number of adults were emerged from control (30.58).

After 120 days of storage, neem oil nano-emulsion exhibited least number of adults (11.90) emerged. Karanj oil nano-emulsion, castor oil nano-emulsion and neem oil exhibited adult emergence values of 15.92, 17.81 and 20.25, respectively which were remained at par with each other.Karanj oil (25.30), eucalyptus oil nano-emulsion (26.32),castor oil (28.62) and eucalyptus oil (29.59)were found at par with each other. The significant higher number of adults were emerged from control (78.32).

After 150 days of storage, least number of adults (31.25) were emerged from neem oil nano-emulsion followed by karanj oil nano-emulsion (38.18) which significantly differed from each other. Castor oil nano-emulsion and neem oil treated seeds recorded adult emergence values of 44.62 and 50.55, respectively which were remained at par with each other followed by karanj oil(57.30) and eucalyptus oil nano-emulsion (59.91). Castor oil (68.23) and eucalyptus oil (70.39)were found at par with each other. The significant higher number of adults were emerged from control (247.12).

After 180 days of storage, the least number of adults (40.20) were emerged from neem oil nano-emulsion followed by karanj oil nano-emulsion (45.29)both significantly differed from each other.Castor oil nano-emulsion was exhibited adult emergence value of 52.56 which followed by neem oil(60.22). Karanj oil(64.96) and eucalyptus oil nano-emulsion (68.89) were found at par with each other. The higher population growth of *R. dominica* among oils and nano-emulsions was found from castor oil (75.17) and eucalyptus oil (82.81). Significantly higher number of adults were emerged from control (350.44).

Earlier, none of the scientists have worked on evaluation of various nano-emulsions as grain protectants based on population growth of *R. dominica* on wheat and therefore, only oils and their efficacy could be compared withwork by earlier researchers. The present findings are in conformity with Nema and Bhargava (2012), Kumawat and Naga (2013), Chintala (2015) and Singh *et al.* (2016) who reported significant reduction in adult emergence of *R. dominica*on stored wheat under oil treatments. However, the number of adults emerged recorded from present studies were found greater in comparison with results obtained by Nema and Bhargava (2012), Chintala (2015), Kumawat & Naga (2013) and lower when compared with Singh *et al.* (2016). This variation could be due to change in variety used and location.

### 4.3.4 Evaluation Based on Reduction in Seed Germination

The data presented in Table 4.11 indicated the percent germination of treated (with oils and nano-emulsions) seeds after six months of storage along with reduction in germination due to the effect of oils and nano-emulsions on germination of wheat.

The data of germination percentages after six months of storage revealed a range of 61.35% to 87.34 %, with an average germination rate of 74.57 %. Significant difference in germination was found among different treatments. The highest germination percentage (87.34) was recorded from control treatment which was significantly differed from other treatments. The next highest germination percentage was recorded from eucalyptus oil nano-emulsion (80.34) which was at par with karanj oil nano-emulsion (80.02) and castor oil nano-emulsion (76.68). Eucalyptus oil treatment exhibited 76.69 % germination. Castor oil(68.34 %) and karanj oil (68.02 %) were remained at par with each other. Neem oil was resulted in least germination percentage (61.35) among various oils and nano-emulsion treatments.

The per cent germination loss due to treatment by oils and nano-emulsions after six months of storage revealed that, among oils and nano-emulsions, lowest per cent reduction in germination was observed from eucalyptus oil nano-emulsion (7.98). Further, the germination loss increased in the order; karanj oil nano-emulsion (8.38 %)<castor oil nano-emulsion (8.74 %)<Eucalyptus oil (12.16 %) <karanj oil (22.10 %)< castor oil (21.74 %). The highest germination loss (29.73 %) was recorded from neem oil.

Average loss in germination due to oils and nano-emulsions was 15.99 %. Average loss in germination due to oils (21.43 %) was higher than that of nano-emulsions (10.55 %). This may point towards reduced impact of nano-emulsions on vital properties of seeds when compared to their corresponding oils and hence nano-emulsions could be a better alternative to oils for grain protection.

Scanty information is available in the published literature regarding germination losses due to effect of nano-emulsions. The present investigation revealed reduction in germination due to effect of oils coincide with studies by Klein and Hebbe (1995) and El-Naby *et al.* (2020) who reported reduction in germination of wheat seeds due to effect of oil treatments. Whereas, Kumawat and Naga (2013), Chintala (2015) and Singh *et al.* (2016) who reported only negligible effect of oils on germination. The variation in germination of wheat seeds due to effect of oils among previous works could be due to change in variety of wheat and method of germination test.

**Table 4.10:Effectiveness of oils and nano-emulsions against *R. dominica* based on population growth**

| | | | Number of adults emerged after indicated days of storage | | | | | |
|---|---|---|---|---|---|---|---|---|
| **Tr. No.** | **Treatments** | **Dose (ml/100 g seeds)** | **30** | **60** | **90** | **120** | **150** | **180** |
| T1 | Castor oil 100 % | 1 | $1.82^{c}$ (3.31) | $2.51^{bc}$ (6.30) | $3.55^{bc}$ (12.60) | $5.35^{b}$ (28.62) | $8.26^{b}$ (68.23) | $8.67^{c}$ (75.17) |
| T2 | Neem oil 100 % | 1 | $1.73^{c}$ (2.99) | $2.44^{cd}$ (5.95) | $3.36^{c}$ (11.29) | $4.50^{cd}$ (20.25) | $7.11^{d}$ (50.55) | $7.76^{e}$ (60.22) |
| T3 | Karanj oil 100 % | 1 | $1.82^{c}$ (3.31) | $2.58^{bc}$ (6.66) | $3.46^{bc}$ (11.97) | $5.03^{bc}$ (25.30) | $7.57^{c}$ (57.30) | $8.06^{d}$ (64.96) |
| T4 | Eucalyptus oil 100 % | 1 | $2.23^{b}$ (4.97) | $2.70^{b}$ (7.29) | $3.78^{b}$ (14.29) | $5.44^{b}$ (29.59) | $8.39^{b}$ (70.39) | $9.10^{b}$ (82.81) |
| T5 | Castor oil nano-emulsion 6 % | 1 | $1.62^{cd}$ (2.62) | $2.23^{d}$ (4.97) | $2.82^{d}$ (7.95) | $4.22^{d}$ (17.81) | $6.68^{d}$ (44.62) | $7.25^{f}$ (52.56) |
| T6 | Neem oil nano-emulsion 6 % | 1 | $1.27^{e}$ (1.61) | $1.62^{f}$ (2.62) | $2.23^{e}$ (4.97) | $3.45^{e}$ (11.90) | $5.59^{f}$ (31.25) | $6.34^{h}$ (40.20) |
| T7 | Karanj oil nano-emulsion 6 % | 1 | $1.41^{de}$ (1.99) | $1.91^{e}$ (3.65) | $2.64^{d}$ (6.97) | $3.99^{d}$ (15.92) | $6.18^{e}$ (38.18) | $6.73^{g}$ (45.29) |
| T8 | Eucalyptus oil nano-emulsion 6 % | 1 | $1.73^{c}$ (2.99) | $2.44^{cd}$ (5.95) | $3.45^{bc}$ (11.90) | $5.13^{b}$ (26.32) | $7.74^{c}$ (59.91) | $8.30^{d}$ (68.89) |
| T9 | Control | 1 | $2.64^{a}$ (6.97) | $4.19^{a}$ (17.56) | $5.53^{a}$ (30.58) | $8.85^{a}$ (78.32) | $15.72^{a}$ (247.12) | $18.72^{a}$ (350.44) |
| **S. Em. ±** | | | 0.09 | 0.09 | 0.10 | 0.16 | 0.13 | 0.08 |
| **F Test (T)** | | | Sig. | Sig. | Sig. | Sig. | Sig. | Sig. |
| **C.V. (%)** | | | 8.81 | 8.81 | 5.35 | 5.58 | 2.89 | 1.55 |

**Notes:** 1. Figures in parentheses are retransformed values of square root transformed values

2. Treatment mean(s) with letter(s) in common are not differed significantly by Duncan's New Multiple Range Test (DNMRT) at 5% level of significance

**Table 4.11: Effect of oils and nano-emulsions against *R. dominica* based on germination of wheat during storage**

| Tr. No. | Treatment | Germination of treated seeds after six months of storage (%) | Loss in germination (%) |
|---|---|---|---|
| T1 | Castor oil 100 % | 55.76[e] (68.34) | 21.74 |
| T2 | Neem oil 100 % | 51.56[f] (61.35) | 29.73 |
| T3 | Karanj oil 100 % | 55.56[e] (68.02) | 22.10 |
| T4 | Eucalyptus oil 100 % | 61.13[c] (76.69) | 12.16 |
| T5 | Castor oil nano-emulsion 6 % | 63.21[b] (76.68) | 8.74 |
| T6 | Neem oil nano-emulsion 6 % | 58.28[d] (72.36) | 17.12 |
| T7 | Karanj oil nano-emulsion 6 % | 63.45[b] (80.02) | 8.38 |
| T8 | Eucalyptus oil nano-emulsion 6 % | 63.68[b] (80.34) | 7.98 |
| T9 | Control | 69.16[a] (87.34) | - |
| S. Em. ± | | 0.48 | - |
| F Test (T) | | Sig. | |
| C.V. (%) | | 1.38 | |

**Notes:** 1. Figures in parentheses are retransformed values of arc sine transformed values

2. Treatment mean(s) with letter(s) in common are not differed significantly by Duncan's New Multiple Range Test (DNMRT) at 5% level of significance

## Chapter- V

**Summary and Conclusion**

The investigations on "Bionomics of lesser grain borer, *Rhyzopertha dominica* Fabricius and its management through nano-emulsions in stored wheat" were carried out at Seed Laboratory of Regional Research Station, Anand Agricultural University, Anand during 2022-2023.The results drawn from the investigations are summarized below.

## 5.1 BIONOMICS OF LESSER GRAIN BORER ON STORED WHEAT

The biology of the lesser grain borer, *R.dominica* on stored wheat revealed that the pest had gone through four stages *i.e.,* egg, larva, pupa, and adult. The adults of *R. dominica* laid oblong, white and opaque eggs in group or singly among the mixture of damaged grain and frass, at bottom of the storage container and rarely, singly on grain surface. The incubation period and egg hatching of these insect-pests eggs ranged from 5 to 6 days with an average of 5.40 ± 0.50days and 80 per cent, respectively.

The larvae of *R. dominica* passed through four instars and fed inside the grain. The larvae entered the grain after hatching by creating tiny bored holes. Once entered, it started feeding on kernel and embryo from inside without no conspicuous sign visible outside the grain. The first instar larvae appeared tiny, translucent, whitish with hairy body and brownish head and its duration was varied from 4 to 6 days with an average 4.30 ± 0.72days. The length of larvae was varied from 0.68 to 0.75 mm with an average of 0.71 ± 0.01 mm and breadth was varied from 0.15 to 0.19 mm with an average of 0.18 ± 0.00 mm. The second instar larvae were marked by an increased size, reduced translucency, hairy whitish body, improved visibility of abdominal segmentation except at the posterior portion which remained voluminous and lustrous. Its duration was varied from 3 to 4 days with an average 3.50 ± 0.31days. The length and breadth were varied from 1.10 to 1.40 mm with an average of 1.22 ± 0.06 mm and 0.36 to 0.47 mm with an average of 0.40 ± 0.02 mm, respectively. The third instar larvae were scarabeiform (C- shaped), opaque, with white robust body and marked by clear body segmentation with hairs. The duration was ranged from 4 to 5 days with an average 4.60 ± 0.56days, whereas the length and breadth were varied from 1.98 to 2.34 mm with an average of 2.18 ± 0.06 mm and 0.56to 0.75 mm with an average of

0.63 ± 0.04 mm, respectively. The fourth instar larvae were similar to the previous instar, but with an enhanced size and pigmentation of head and mandibles. The duration was varied from 4 to 8 days with an average 6.80 ± 0.92days. The length and breadth were varied from 2.58 to 2.72 mm with an average of 2.68 ± 0.02mm and 0.89 to 1.03 mm with an average of 0.98 ± 0.02 mm, respectively.

The prepupal duration was one day. However, the pupal period was varied from 3 to 6 days with an average of 3.85 ± 0.87days. The adult had a dark brown body with length ranging from 2.36 to 2.54 cm with average of 2.45 ± 0.03 and breadth ranging from 0.84 to 0.90 cm with an average of0.88 ± 0.01. The elytra of adults were longitudinally pitted, pronotum was hood like, covering the head region and antenna was clubbed with ten segments. The sex of adults was determined based on presence of pale-yellow colour on the last few abdominal segments of female. Thesex ratio was found to be 1 : 1(male: female).The average period of pre-oviposition was 2.5 ± 0.60 days with a minimum and maximum pre-oviposition period of 2 to 4 days, respectively. The average period of oviposition was 22.50 ± 2.23 days with a minimum and maximum oviposition period of 20 to 25, respectively. The average period of post-oviposition was 6.85 ± 1.32 days with a minimum and maximum post-oviposition period of 5 to 9, respectively. The total male and female lifespan were varied from 49 to 67 days and 50 to 70 days, respectively with an average male and female lifespan of59.93 ± 3.86 and 63.55 ± 4.88days, respectively. The fecundity was varied from 120 to 250 eggs with an average 191 ± 40.5eggs.

## 5.2 SCREENING OF DIFFERENT VARIETIES OF WHEAT FOR SUSCEPTIBILITY TO LESSER GRAIN BORER, *R.dominica*

On the basis of population growth of *R. dominica*, the lower number of adults were emerged in the variety, GW513and HI1544which found resistant to this pest. Whereas,GW206, GW1 and LOK1were found moderately resistant. GDW 1255, GADW3, GW499 and GW496were found moderately susceptible. GW451showed the highest adult emergence and considered as susceptible against *R. dominica*.

In terms of per cent weight loss, the variety GW513was categorized into resistant group, whereas, varieties, GW 206, HI 1544, GW 1 and LOK 1were grouped under moderately resistant category. The varieties, GW 451, GW 496, GDW 1255 and GW 499were considered as moderately susceptible andGADW3was found highly susceptible.

Based on per cent reduction in germination after six months release of *R. dominica*, GW513 and GW206 varieties recorded lower per cent germination loss and categorized as resistant variety. GDW 1255, GADW3, GW1 and GW496 were found under moderately resistant category. GW499 and LOK1 were emerged out as moderately susceptible to *R. dominica*. Varieties GW451 and HI1544were found under susceptible category in respect of germination reduction.

There was positive correlation between all the different physical characters under investigation and weight loss of grain due to infestation by *R. dominica*. Seed weight ($r = 0.87^{**}$) and length ($r = 0.73^{*}$)were positively correlated toper cent weight loss. Seed hardness was negatively correlated with weight loss ($r = - 0.48$),whereas it was positively and non-significantly correlated with seed width($r = 0.61$).

From the point of view of categorization, it can be concluded that the variety,GW513 was found resistant based on germination loss, population growth and weight loss.GW 1 variety was found moderately resistant based on all the three parameters. The variety, GDW 1255 was moderately susceptible based on population growth and weight loss, whereas it was moderately resistant based on germination loss. The variety,LOK1 was found moderately susceptible based on germination loss, whereas it was moderately resistant based on weight loss and population growth. The variety,GW451 was found susceptible based on population growth and germination loss, whereas it was moderately susceptible based on weight loss. Both GW496 and GADW3 were found moderately susceptible based on population growth, whereas it was moderately resistant based on germination loss. The variety,HI1544 was found resistant based on population growth, moderately resistant based on weight loss and susceptible based on germination loss. The variety,GW206 was found moderately resistant based on population growth and weight loss, whereas it was resistant based on germination loss.

## 5.3 EVALUATION OF DIFFERENT NON-EDIBLE OILS AND ITS NANO EMULSIONS AS GRAIN PROTECTANTS AGAINST LESSER GRAIN BORER, *R. dominica*ON STORED WHEAT

All the prepared nano-emulsions found within the nano-scale (1 – 999 nm).Zeta potential, kilo count per second and polydispersity indexwere also with in a stable range. The hydrodynamic size of the castor oil, neem oil, karanj oil and eucalyptus oil nano-emulsions was found to be 64.07 ± 1.32 nm, 37.04 ± 0.75 nm, 33.90 ± 0.29 nm and 46.56 ± 1.25 nm, respectively. The zeta potential of the nano-emulsions was -41.00± 6.45 mV, -53.90± 6.30 mV, 54.90± 6.25 mV, and -30.90 ± 6.82 mV for castor oil, neem oil, karanj oil and eucalyptus oil, respectively. The kilo count per second was measured as 291.30 ± 1.20, 297.80 ± 1.40, 271.70 ± 0.30 and 305.60 ± 3.10 for castor oil, neem oil, karanj oil and eucalyptus oil, respectively. The polydispersity index values were determined as 0.318 ± 0.07, 0.397 ± 0.02, 0.345 ± 0.01 and 0.390 ± 0.02 for castor oil, neem oil, karanj oil and eucalyptus oil, respectively.

Overall, it can be concluded that among oils and nano-emulsions tested at 1 % (v/w), neem oil nano-emulsion (46.91%) evolved to be the most effective treatment exhibiting highest mean adult mortality. Karanjoil nano-emulsion (42.4 %), castor oil nano-emulsion (31.35 %), neem oil (21.68 %)andkaranj oil(20.41 %) were found as next effective treatments. Eucalyptus oil nano-emulsion(16.76 %) and castor oil(14.73 %) were equally effective to cause mortality to *R. dominica*.Eucalyptus oil(9.72 %) was the least effective among the tested oils and nano-emulsions in terms of adult mortality, however, all the nano-emulsions and oils were found effective as grain protectants in terms of adult mortality when compared to control. All nano-emulsions were found more effective than the oils, except eucalyptus oil nano-emulsion, however nano-emulsions were more effective than their corresponding oils.

Based on the population growth of *R. dominica* aftersix months of storage, the least number of adults were recorded in neem oil nano-emulsion (40.20) followed by other treatments in the order: karanj oil nano-emulsion (45.29)> castor oil nano-emulsion(52.56)> neem oil (60.22)>karanj oil (64.96)> eucalyptus oil nano-emulsion(68.89)> castor oil(75.17)> eucalyptus oil(82.21)> control (350.44).

The per cent germination loss due to treatment by oils and nano-emulsions after six months of storage revealed that, among oils and nano-emulsions, lowest per cent reduction in germination was observed from eucalyptus oil nano-emulsion (7.98). Further, the germination loss increased in the order; karanj oil nano-emulsion (8.38 %) < castor oil nano-emulsion (8.74 %) < Eucalyptus oil (12.16 %) <karanj oil (22.10

%) < castor oil (21.74 %). The highest germination loss (29.73 %) was reported from neem oil. The average loss in germination due to oils (21.43 %) was higher than that of nano-emulsions (10.55 %).

Thus, in a nutshell both larval and adult stages of *R. dominica* damaged stored wheat due to its ability to feed and multiply gregariously. Identifying and adopting varieties with pest resistance characteristics and use of plant oils could help mitigate the risk of grain damage. Nano-emulsions which exhibited enhanced effectiveness even at lower concentrations in present studies highlighted the need for further investigations on toxicity profiles and determining the safe and effective concentrations for application as next-generation prophylactic grain protectants.

**Bibiliography**


Abbott, W. S. (1925). A method of computing the effectiveness of an insecticide. *Journal of Economic Entomology*, 18 (2), 265-267.

Abouelatta, A. M., Khalil, F. M. A., & Arab, R. B. A. (2022). Study on *Tagetes minuta* (Fam. Asteraceae) essential oil and its effects on some stored product insects. *Agricultural Mechanization in Asia, Africa & Latin America* 53 (12), 11191 – 11203.

Adak, T., Barik, N., Patil, N. B., Gadratagi, B. G., Annamalai, M., Mukherjee, A. K., & Rath, P. C. (2020). Nanoemulsion of eucalyptus oil: An alternative to synthetic pesticides against two major storage insects (*Sitophilus oryzae* and *Tribolium castaneum*) of rice. *Industrial Crops and Products*, 143, 111849.

Adedire, C. O. (2001). Biology, ecology and control of insect pests of stored grains. *Dave Collins Publication*, pp. 59-94.

Afshan, D. (2005). Population development of red flour beetle, *Tribolium castanium* (Herbst.) and lesser grain borer, *Rhizopertha dominica* (Fab.) on wheat grain (Abadgar). *National Agricultural Research Centre Documentation,* 50.

Ajaykumara, K. M., Thirumalaraju, G. T., & Anjali, A. S. (2018). Seasonal variations in the biology of lesser grain borer *Rhyzopertha dominica* (F.) (Coleoptera: Bostrychidae) on stored maize under laboratory conditions. *Journal of Entomology and Zoology Studies*, 6 (1), 516-522.

Anonymous (2014). United nations, food and agriculture organization, statistics division (FAOSTAT). Retrieved from https://www.fao.org/faostat/en/#data/QC

Anonymous (2020). Selected state-wise area, production and productivity of wheat in India. *Indiastat.com.* Retrived from https://www.indiastat.com/table/Gujarat-state/agriculture/selected-state-wise-area-production-productivity-w/1326556

Anonymous (2022). Third advanced estimate of area, production and yield of major *kharif/rabi*/summer crops of Gujarat state for the year 2022-2023. Retrieved from https://dag.gujarat.gov.in/estimate.html

Atwal, A. S. (1994). Insect pest of stored grain and other products: agricultural pest of India and south east asia, *National Book Foundation*, 2, pp. 402-405.

Awadalla, S., El-Serafi, H. A., Ghanim, A. A., & Shalaby, M. M. (2023). The effect of different host grains on the biological aspects of the lesser grain borer *Rhyzopertha dominica* (Fab.) (Bostrichidae: Coleoptera). *Journal of Plant Protection and Pathology*, 14 (4), 105 -108.

Baghla, K., Rathore, B. S., Saxena, S., Shivhare, R., & Tetarwal, J. M. (2023). Varietal screening of wheat genotype against, *Rhyzopertha dominica* Fab. (Lesser grain borer). *The Pharma Innovation Journal*, 12 (2), 3742-3748.

Bashir, T. (2000). Pheromone communication and host-finding behaviour of *Rhyzopertha dominica* (F.) (Coleoptera: Bostrichidae) (Doctoral dissertation, University of Greenwich).

Bhargava, M. C., & Akhter, H. (2014). Relative susceptibility of different varieties of wheat to lesser grain borer, *Rhyzopertha dominica*. *Journal of Insect Science (Ludhiana)*, 27 (2), 189-194.

Broughton, M. J., & Dunkel, F. V. (2006). Interactions of wheat variety, production environments, and prior insect damage on postharvest resistance to the lesser grain borer. *Journal of Economic Entomology*, 99 (5), 1826-1834.

Burges, H.D. (2008). Development of the khaprabeetle, *Trogoderma granarium* (Everts) in the lower part of its temperature range. *Journal of Stored Products Research*, 44 (1), 32-35.

Campbell, T. L. M. (2002). Thermal biology of the lesser grain borer *Rhyzopertha dominica* (F.) (Bostrichidae) and the warehouse pirate bug *Xylocoris flavipes* (Reuter) (Anthocoridae) (Doctoral dissertation, Montana State University-Bozeman).

Campos, E. V., De Oliveira, J. L., Pascoli, M., De Lima, R., & Fraceto, L. F. (2016). Neem oil and crop protection: from now to the future. *Frontiers in Plant Science*, 7, 1494.

Chandel, R. K., Nebapure, S. M., Sharma, M., Subramanian, S., Srivastava, C., & Khurana, S. M. (2021). Insecticidal and repellent activities of eucalyptus oil against lesser grain borer *Rhyzopertha dominica* (Fabricius). *Journal of Microbiology, Biotechnology and Food Sciences*, 9 (3), 525-529.

Chintala, S. (2015). Biology, varietal susceptibility and management of lesser grain borer *Rhyzopertha dominica* in stored wheat. (Master's thesis, Junagdh Agricultural University, Junagdh).

Chintala, S., Parekh, K. & Virani, V. R. (2017). Biology and behaviour of lesser grain borer, *Rhyzopertha dominica* (Fabricious) (Coleoptera: Bostrichidae) on stored wheat in laboratory conditions. *Agriculture Update*, 12 (5), 1332-1338.

Chintala, S. & Virani, V.R. (2018). Biology and behaviour of lesser grain borer, *Rhyzopertha dominica* (Fabricious) (Coleoptera: Bostrichidae) on stored wheat in laboratory conditions. *International Journal of Agriculture Sciences*, 10 (5), 5231-5234.

Deshwal, R., Gupta, P. K., Vaibhav, V., Kumar, N., & Kumar, A. (2018). Biology of lesser grain borer (*Rhyzopertha dominica* Fab.) under different temperature and humidity at laboratory condition. *Journal of Entomology and Zoology Studies*, 6 (4), 364-368.

Dinesh, L., & Raj, D. V. (2012). Efficacy of application of four vegetable oils as grain protectant against the growth and development of *Callosobruchus maculatus* and on its damage. *Advances in Bio Research*, 3 (2), 55-59.

Edde, P. A. (2012). A review of the biology and control of *Rhyzopertha dominica* (Fab.) the lesser grain borer. *Journal of Stored Product Research*, 48, 1-18.

Edde, P. A., & Phillips, T. W. (2006). Longevity and pheromone output in stored-product Bostrichidae. *Bulletin of Entomological Research*, 96 (6), 547-554.

Elango, G., Roopan, S. M., Dhamodaran, K. I., Elumalai, K., Al-Dhabi, N. A., & Arasu, M. V. (2016). Spectroscopic investigation of biosynthesized nickel nanoparticles and its larvicidal, pesticidal activities. *Journal of Photochemistry and Photobiology*, 162, 162-167.

El-Naby, A., Shimaa, S. I., Mahmoud, F. H., El-Habal, N. A., Abdou, M. S., & Abdel-Rheim, K. H. (2020). Efficacy of prepared castor oil nanoemulsion formulation against rice weevil *sitophilus oryzae* on stored wheat grain and its acute toxic effect on albino rats. *Egyptian Scientific Journal of Pesticides,* 6 (4), 33 – 44.

El-Sheikh, T. A. A., Hassanein, A. A., Radwan, E. M. M., & Abo-Yousef, H. M. (2005). Biochemical effects of certain plant oils on the lesser grain borer, *Rhyzopertha dominica* (F.). *Annals of Agricultural Science*, 50 (2), 729-737.

Eyidozehi, K., Ravan, S., Khani, A., & Khormali, S. (2014). Amount damage of Lesser grain borer on Viability and physical characteristics of seed wheat varieties in laboratory condition in Golestan. *Journal of Novel Researches on Plant Protection*, 6 (2), 172-163.

Fernandez, P., André, V., Rieger, J., & Kühnle, A. (2004). Nano-emulsion formation by emulsion phase inversion. *Colloids and Surfaces A: Physicochemical and Engineering Aspects*, 251 (1-3), 53-58.

Gami, P. S., Patel, J. J., Pansara, Z. H., Patel, K. G., & Thakar, P. K. (2022). Susceptibility of different wheat varieties/genotypes to lesser grain borer, *Rhyzopertha dominica* (Fabricius). *The Pharma Innovation Journal* 11 (5), 1748-1752.

Golden, G., Quinn, E., Shaaya, E., Kostyukovsky, M., & Poverenov, E. (2018). Coarse and nano emulsions for effective delivery of the natural pest control agent pulegone for stored grain protection. *Pest Management Science*, 74 (4), 820-827.

Halstead, D. G. H. (1963). External sex differences in stored-products coleoptera. *Bulletin of Entomological Research*, 54 (1), 119-134.

Hashem, A. S., Awadalla, S. S., Zayed, G. M., Maggi, F., & Benelli, G. (2018). *Pimpinella anisum* essential oil nanoemulsions against *Tribolium castaneum*, insecticidal activity and mode of action. *Environmental Science and Pollution Research*, 25 (19), 18802-18812.

Ibrahim, S. S. (2022). Polyethylene glycol nanocapsules containing *Syzygium aromaticum* essential oil for the management of lesser grain borer, *Rhyzopertha dominica*. *Food Biophysics*, pp. 1-12.

Ileke, K. D., & Oni, M. O. (2011). Toxicity of some plant powders to maize weevil [*Sitophilus zeamais* (motschulsky) (Coleoptera: Curculiondae) on stored wheat grains (*Triticum aestivum*). *African Journal of Agricultural Research*, 6 (13), 3043-3048.

International Seed Testing Association (2015). *International rules for seed testing*, 2015. Basserfdorf, Switzerland.

Iqbal, H., Jahan, N., Khalil-ur-Rahman, & Jamil, S. (2022). Formulation and characterisation of *Azadirachta indica* nanobiopesticides for ecofriendly control of wheat pest *Tribolium castaneum* and *Rhyzopertha dominica*. *Journal of Microencapsulation*, 39 (7-8), 1-16.

Kalaf, J. M. (2021). Testing the effect of some botanical oils on some aspects of the life performance of lesser grain borer *Rhyzopertha dominica* (F.) (Coleoptera: Bostrichidae). *Annals of the Romanian Society for Cell Biology*, 25(6), 12008-12016.

Keever, D. W. (1983). Distribution patterns of lesser grain borer (Coleoptera: Bostrichidae) in towers of wheat, and effects of the presence of the granary weevil (Coleoptera: Curculionidae). *Journal of Economic Entomology*, 76(3), 492-495.

Khanzada, F. (2007). Population fluctuation and mode of damage of lesser grain borer on different varieties of wheat. *National Agricultural Research Centre Documentation,* 50.

Khemira, S., Mediouni-Ben JemÃ¢ a, J., Haouel, S., & Khouja, M. L. (2012). Repellent activity of essential oil of *Eucalyptus astringens* against *Rhyzopertha dominica* and *Oryzaephilus surinamensis*. *ISHS Acta Horticulturae 997: International Symposium on Medicinal and Aromatic Plants - SIPAM 2012,* 207-213.

Kher, R. H. (2006). Evaluation of vegetable oil and insecticides as grain protectants and varieties for susceptibility against *Rhyzopertha dominica (F.)* infesting wheat under storage condition (Doctoral thesis, Anand Agricultural University, Anand).

Khokhar, N. (2005). Relative resistance of some wheat varieties against lesser grain borer in laboratory. *National Agricultural Research Centre Documentation,* 61.

Klein, J. D., & Hebbe, Y. (1995). Effect of the treatment of wheat seeds with vegetable oils on germination and emergence. *Experimental Agriculture*, 31 (3), 291-298.

Klys, M. (2008). Temperature influence at the feeding activity of lesser grain borer *Rhyzopertha dominica* F. (Coleoptera, Bostrichidae). *Ecological Chemistry and Engineering. A*, 15 (4-5), 355-360.

Korawar, L.H. (2019). Varietal screening and management of *Rhyzopertha dominica* (Fabricius) in stored Wheat. (Master's thesis, Mahatma Phule Krishi Vidyapeeth, Rahuri).

Kucerova, Z., & Stejskal, V. (2008). Differences in egg morphology of the stored-grain pests *Rhyzopertha dominica* and *Prostephanus truncatus* (Coleoptera: Bostrichidae). *Journal of Stored Products Research*, 44 (1), 103-105.

Kumar, P. S., Uma, M. T., & Padmakumari, A. P. (2017). Comparative biology of *Rhyzopertha dominica* (Fab.) in major cereals. *International Journal of Current Microbiological Application Science*, 6 (7), 2205-2210.

Kumawat, K. C. (2007). Effect of abiotic factors on biology of *Rhyzopertha dominica* (fab.) on wheat. *Annals of Plant Protection Sciences*, 15 (1), 111-115.

Kumawat, K. C., & Naga, B. L. (2013). Effect of plant oils on the infestation of *Rhyzopertha dominica* (Fab.) in wheat, *Triticum aestivum* Linn. *Journal of Plant Protection Research*, 53 (3), 301-304.

Kumawat, K., Kumar, R., Rana, B. S., & Verma, A. (2022). To screen wheat varieties for their resistance to *Rhyzopertha dominica. The Pharma Innovation Journal,* 11 (2), 954-957.

Mark, A. C., Severtson, D. L., Brumley, C. J., Szito, A., Foottit, R. G., Grimm, M., Munyard, K. & Groth, D. M. (2010). A rapid non-destructive DNA extraction method for insects and other arthropods. *Journal of Asia-Pacific Entomology*, 13 (3), 243-48.

Mason, L. J., & McDonough, M. (2012). Biology, behavior and ecology of stored grain and legume insects. *Stored Product Protection*, 1 (7), 11-12.

McGaughey, W. H., Speirs, R. D., & Martin, C. R. (1990). Susceptibility of classes of wheat grown in the United States to stored-grain insects. *Journal of Economic Entomology*, 83 (3), 1122-1127.

Meenakshi, U., & Srivastava, A. K. (2010). Response of wheat varieties to lesser grain borer, *Rhyzopertha dominica* (Fabr.). *Annals of Plant Protection Sciences*, 18 (1), 144-147.

Mehta, V. (2020). Impact of different wheat cultivars on lesser grain borer, *Rhyzopertha dominica* (F.) (Coleoptera: Bostrichidae) under storage conditions. *Journal of Entomology and Zoology Studies*, 8 (3), 1063-1066.

Mumm, T. L., O'Neill, K. M., & Weaver, D. K. (2001). Thermal biology of the lesser grain borer, *Rhyzopertha dominica* (F.) (Coleoptera: Bostrichidae). *The ESA 2001 Annual Meeting - 2001: An Entomological Odyssey of ESA,* D0076.

Naik, R. H., Mohankumar, S., Naik, S. O., Pallavi, M. S., Srinivasan, M. R., & Chandrasekaran, S. (2016). Influence of food sources on developmental period of *Rhyzopertha dominica, Tribolium castaneum* and *Sitophilus oryzae*. *Indian Journal of Plant Protection*, 1 (44), 63-68.

Natanam, R., Kadirvel, R., & Chandrasekaran, D. (1989). Chemical composition of karanja (*Pongamia glabra Vent*) kernel and cake as animal feed. *Indian Journal of Animal Nutrition*, 6 (3), 270-273.

Nema, R., & Bhargava, M. C. (2012). Management of *Rhyzopertha dominica* (Fabricius) on wheat with essential oils. *Journal of Insect Science (Ludhiana)*, 25 (4), 388-390.

Obeng- Ofori, D. (1995). Plant oils as grain protectants against infestations of *Cryptolestes pusillus* and *Rhyzopertha dominica* in stored grain. *Entomologia Experimentalis et Applicata*, 77 (2), 133-139.

Obretenchev, D., & Staneva, E. (2012). Tolerance of soft wheat varieties selected in Bulgaria to the lesser grain borer *Rhyzopertha dominica* F. (Coleoptera: Bostrychidae) during storage. *Acta Entomologica Bulgarica*, 15 (1), 65-74.

Pal, R. K., Dwivedi, R. K., & Singh, D. K. (2011). Effect of lesser grain borer (*Rhyzopertha dominica* Fab.) on wheat seeds during storage. *Current Advances in Agricultural Sciences*. 3 (1), 69-69.

*Patel, I. S., Prajapati, B.G., Patel, G. M., & Pathak, A. R. (2002). Response of castor genotypes to castor semilooper, *Achaea janata* Fab., *Journal of Oilseeds Research*, 19 (1), 153.

Perdomo, F. A., Acosta-Osorio, A. A., Herrera, G., Vasco-Leal, J. F., Mosquera-Artamonov, J. D., Millan-Malo, B., & Rodriguez-Garcia, M. E. (2013). Physicochemical characterization of seven mexican *Ricinus communis* L. seeds & oil contents. *Biomass and Bioenergy*, 48, 17-24.

Phillips, T. W., & Throne, J. E. (2010). Biorational approaches to managing stored-product insects. *Annual Review of Entomology*, 55 (1), 375-397.

Pruthi, H. S. & Singh, M. (1950). Pests of stored grain and their control. *Indian Journal of Agricultural Sciences*, 18 (4), 52-58.

Rao, H. G., & Wilbur, D. A. (1972). Loss of wheat weight from feeding of lesser grain borer. *Journal of the Kansas Entomological Society*, 238-241.

Sabbour, M. M., & El-Aziz, S. E. A. (2016). Efficacy of three essential oils and their nano-particles against *Sitophilus granarius* under laboratory and store conditions. *Journal of Entomological Research*, 40 (3), 229-234.

Sarin, K., & Sharma, K. (1983). Study of antibiosis in wheat varieties part I. Correlation of diapause and growth index. *Bulletin of Grain Technology*, 21, 24-30.

Sarkar, S., Gil, J. D. B., Keeley, J., & Jansen, K. (2021). *The use of pesticides in developing countries and their impact on health and the right to food.* Wageningen, Netherlands: Berlin.

Sayed, T. S., Hirad, F. Y., & Abro, G. H. (2006). Resistance of different stored wheat varieties to Khapra beetle, *Trogoderma granarium* (Everest) and lesser grain borer, *Rhizopertha dominica* (Fabricus). *Pakistan Journal of Biological Sciences*, 9 (8), 1567-1571.

Scheff, D. S., Campbell, J. F., & Arthur, F. H. (2021). Seasonal, landscape, and attractant effects on lesser grain borer, *Rhyzopertha dominica* (F.), captures in Northeast Kansas. *Agronomy*, 12 (1), 99.

Seada, M. A., & Hamza, A. M. (2023). Comparative morphology of sensilla of antennae, maxillary and labial palpi of adult *Rhyzopertha dominica* (F.)

(Coleoptera: Bostrichidae), with specific reference to the typology and possible functions. *The Journal of Basic and Applied Zoology*, 84 (1), 14.

Shahbandeh, M. (2023). Wheat: production volume worldwide 1990/1991-2022/2023. Retrieved from https://www.statista.com/statistics/267268/production-of-wheat-worldwide-since-1990/

Sharma, S., Loach, N., Gupta, S., & Mohan, L. (2020). Phyto-nanoemulsion: An emerging nano-insecticidal formulation. *Environmental Nanotechnology, Monitoring & Management*, 14, 100331.

Shukla, S., Dubey, A., & Chandel, B. S (2020). Varietal resistance of certain indigenous wheat, *Triticum aestivum* (Linn.) for the larval development of lesser grain borer, *Rhyzopertha dominica* (Coleoptera: Bostrichidae). *International Journal of Fauna and Biological Studies,* 7 (4), 126-131.

Singh, S., & Sharma, D. K. (2021). Screening of wheat varieties against lesser grain borer *Rhyzopertha dominica. Indian Journal of Entomology,* 83 (4), 602-605.

Singh, S., Sharma, D. K., & Gill, R. S. (2016). Evaluation of three plant oils for the control of lesser grain borer, *Rhyzopertha dominica* (Fabricius) in stored wheat. *Journal of Insect Science*, 29 (1), 162-169.

Srivastava, P. K., Tripathi, B. P.,Girish, G. K. and Krishnamurthy, K. (1973). Studies on the assessment of losses: Conventional grain storage practices and losses in rural areas in western U.P. *Bulletin of Grain Technology*, 11 (2), 129-139.

Talukder, F. A. (2006). Plant products as potential stored-product insect stored management agents-a mini review. *Emirates Journal of Food and Agriculture*, 18 (1), 17-32.

Tandon, S., Mittal, A. K., & Pant, A. K. (2008). Evaluation of essential oils from some aromatic plants against pulse beetle (*Callosobruchus chinensis* L.) control. *Journal of Food Legumes*, 21 (4), 281-282.

Toews, M. D., Cuperus, G. W., & Phillips, T. W. (2000). Susceptibility of eight US wheat cultivars to infestation by *Rhyzopertha dominica* (Coleoptera: Bostrichidae). *Environmental Entomology*, 29 (2), 250-255.

Win, N., & Rolania, K. (2020). Influence of seasonal variation on the biology of lesser grain borer, *Rhyzopertha dominica*, (Fabricius) on wheat. *Journal of Entomology and Zoology Studies*, 8 (5), 285-290.

*Originals not seen